ENCOUNTERING LIFE IN THE UNIVERSE

Encountering Life in the Universe

Ethical Foundations and Social Implications of Astrobiology

CHRIS IMPEY

ANNA H. SPITZ

WILLIAM STOEGER

Editors

THE UNIVERSITY OF ARIZONA PRESS

TUCSON

The University of Arizona Press

www.uapress.arizona.edu

Library of Congress Cataloging-in-Publication Data
Encountering life in the universe : ethical foundations and social implications of astrobiology / Chris Impey, Anna H. Spitz, William Stoeger, editors.
p. cm.
"This book grew out of the workshop 'Astrobiology: Expanding our Views of Society and Self,' held at the University of Arizona's (UA) Biosphere 2 Institute in May 2008."
Includes bibliographical references and index.
ISBN 978-0-8165-2870-7 (pbk. : alk. paper)
1. Exobiology—Moral and ethical aspects. 2. Exobiology—Social aspects. 3. Life on other planets—Moral and ethical aspects. 4. Life on other planets—Social aspects. I. Impey, Chris. II. Spitz, Anna H., 1954– III. Stoeger, William R.
QH326.E537 2012
576.8'39—dc23

2013011212

Publication of this book is made possible in part by funding from the University of Arizona's Steward Observatory and the Vatican Observatory.

Manufactured in the United States of America on acid-free, archival-quality paper containing a minimum of 30% post-consumer waste and processed chlorine free.

18 17 16 15 14 13 6 5 4 3 2 1

Table of Contents

Illustrations

Figures

Table

Preface

Astrobiology is the study of life's relationship to the rest of the cosmos. Its major themes include the origin of life and its precursor materials, the evolution of life on Earth, its future prospects on and off the Earth, and the occurrence of life elsewhere. Behind each of these themes is a multi-disciplinary set of questions involving physics, chemistry, biology, geology, astronomy, planetology, and other fields, each of which connects more or less strongly to the central questions of astrobiology. Stimulated by new capabilities for scientific exploration on and off the Earth, astrobiology is establishing itself as a distinct scientific endeavor.

To what extent does progress in astrobiology stimulate social, cultural, and ethical issues? The grand questions, such as whether we are alone in the cosmos, or what would happen to humankind were we to encounter an advanced extraterrestrial civilization, have been debated in one form or another for centuries. But the issues are becoming more focused and more imperative as the science advances. To what extent should we permit ourselves to profoundly modify alien landscapes—from the surface of the Moon to the nuclei of comets—in search of the answers to scientific questions? Is exploration of Mars worth the risk of exposing humankind to potentially novel and lethal organisms that might be returned in samples? Equally, is such exploration worth the risk of contaminating Mars—or any other potentially inhabited planetary body—with terrestrial organisms that might be lethal to indigenous life? Do we have the right to seed life on previously uninhabited worlds? If there exists life on Earth whose origin is demonstrably separate from our own, do we have a moral obligation to protect it? Should governments restrict the rights of individuals to broadcast messages into space if and when artificially generated radio signals from another planetary system are discovered? Is it ethical to create an artificial life-form in seeking to understand the origin of life? How might scientific findings about the commonality or singularity of Earth as an inhabited world affect the outlook of diverse human cultures? How do we determine and respond to the moral status of extraterrestrial life?

This collection of essays presents in a single volume the key social, cultural, and ethical issues raised by astrobiological research. Its treatment of the cultural issues is remarkably broad. Contributions range from philosophical questions about the origin of self-organization in the cosmos to "practical" ethical issues associated with cross-contamination of Earth and other bodies in our Solar System. The collection grew out of a workshop, Astrobiology: Expanding Our Views of Society and Self, held in May 2008 at the University of Arizona's Biosphere 2 Institute. There, invited experts deliberated in the shadow of the Biosphere 2 artificial ecosystem facility, surrounded by the high Sonoran Desert and the sky islands of southern Arizona—one of the world's most diverse and beautiful ecosystems.

At the conclusion of the workshop the participants agreed on a joint statement regarding the need for international consideration of the basis for planetary protection, specifically to broaden the basis of planetary protection from simply the preservation of future science exploration to consideration of ethical issues raised by possible alternative types of life. Attendees signed and submitted the resolution to the Committee on Space Research (COSPAR) Planetary Protection Panel of the International Council for Science. In 2010 the COSPAR Bureau held an international workshop to discuss ethical issues and planetary protection.

This volume reflects novel insights into, and new connections between, scientific research in astrobiology and ethical, social, and philosophical problems. Its chapters will serve as an important resource for thoughtful solutions to the ethical, social, and philosophical challenges that will inevitably arise from further advances in astrobiological scientific research.

Jonathan I. Lunine
Anna H. Spitz

Acknowledgments

This book grew out of the workshop "Astrobiology: Expanding Our Views of Society and Self," held at the Biosphere 2 Institute, University of Arizona (UA), in May 2008. Members of the organizing committee were:

- Travis Huxman, Biosphere 2 Director, UA Department of Ecology and Evolutionary Biology
- Chris Impey, Professor, UA Department of Astronomy
- Tom Lindell, Emeritus Professor, UA Department of Molecular and Cellular Biology
- Anna Spitz, Program Manager, UA College of Science Center for Astrobiology
- William Stoeger, Vatican Observatory
- Nick Woolf, Professor, UA Department of Astronomy
- Education and Public Outreach: Julia Olsen, Research Assistant, UA Department of Astronomy
- Logistics: Cathi Duncan, LAPLACE Program Coordinator, UA Department of Astronomy

The editors and authors wish to thank the workshop organizing committee for the inspiration for this volume.

ENCOUNTERING LIFE IN THE UNIVERSE

CHAPTER ONE

Astrobiology, Ethics, and Philosophy

William R. Stoeger
VATICAN OBSERVATORY

Chris Impey
STEWARD OBSERVATORY, UNIVERSITY OF ARIZONA

Anna H. Spitz
LUNAR AND PLANETARY LABORATORY,
UNIVERSITY OF ARIZONA

Motivation, Philosophy, and Vision

The philosophy that guides this volume, *Encountering Life in the Universe: Ethical Foundations and Issues and Social Implications*, is captured in the title of the conference that gave it birth: Astrobiology: Expanding Our Views of Society and Self. It is certainly about the ethical issues raised by astrobiology and by the possibility of life originating and evolving elsewhere in the cosmos. But it is even more about the perspectives, values, and attitudes such knowledge and understanding can engender in us as individuals and as communities, and in society more generally—attitudes toward ourselves in all our cultural and ethnic diversity, toward the other living organisms with whom we share this planet, and toward other possible, and possibly very different, extraterrestrial life-forms—both conscious and preconscious.

This volume brings together, as did the workshop, philosophers and physical scientists who have thought deeply about fundamental issues concerning research in astrobiology. The intellectual journeys of the contributors reveal efforts to understand the science, the implications of the science, the underlying philosophies (and questions) that help place the science in context, and the wider societal concerns about the science and its outcomes. They reveal the efforts of active researchers to expand on their own areas of expertise, in fact to move outside their comfort zones and experience, to understand the larger meaning of their work. In coming together to discuss the issues and make use of the knowledge of experts in different fields, all participants further informed their own efforts. Grappling with the fundamental questions and outcomes of research in astrobiology is vital to creating responsible, thoughtful, and ultimately rigorous science, as well as, one hopes, useful foundations for the larger society, whether students, the general public, or policy makers.

Such informed, open, and appreciative perspectives and attitudes give rise to enlightened ethics: judgments of what is beneficial and good, of what is of value—not just to us but also to the *other*—and of when and how to act or refrain from acting in various circumstances. Just as importantly, such attitudes and perspectives guide how we are to prioritize those values in the policies we pursue as we encounter life—and possibly intelligent life—here and elsewhere in the Universe. Our ethical principles and decisions are based in and supported by this scientific knowledge and understanding along with these crucial attitudes, stances, and values we assume in light of what we know or do not know and what we imagine, fear, and hope for. Some contributors to this volume discuss what we know and understand about our Universe—physically, chemically, and biologically—and how we respond to such knowledge in terms of attitudes and values. Other contributors discuss specific ethical issues that have already arisen or that we foresee will arise.

Exploration of the ethical implications of astrobiology is not as esoteric a pursuit as it might at first appear. It follows directly from the established considerations of other interdisciplinary fields like bioethics and environmental ethics. The advances in biology, biotechnology, and medicine have prompted a new urgency in applying theories of ethics to practical problems such as the boundaries of life. Environmental ethics extends the application of ethics from humans to the nonhuman world using philosophy and ethics to influence how we manage the physical world or interact with other species. Astrobiology moves such applications further to considerations of how we might behave as we explore new worlds or create new life in the laboratory, or interact with extraterrestrial life-forms.

Astrobiology, as Jonathan Lunine and Anna Spitz state in the Preface to this volume, is the interdisciplinary investigation of all the conditions, processes, and stages—astronomical, physical, chemical, geological, environmental—that are necessary for the emergence of life here on Earth and elsewhere in the cosmos. Some of those conditions (such as the expansion and cooling of the Universe and the generation of the chemical elements) are cosmic and originate in deep time, some (such as the characteristics of the nearest star and the potentially life-bearing planet's position within its stellar system) are much more locally astronomical, and others (such as the size, age, and detailed characteristics of the planet and its chemical evolution) are more locally geological and chemical. There may be many necessary conditions for the emergence of life in any astronomical venue, and many possible scenarios for its initiation and development, given different combinations of such conditions. But our current knowledge is inadequate to list combinations of conditions that are sufficient for biology in the general case. In fact, there is scientific and philosophical debate over the basic question of definition—what is life, and how would we recognize it if we found it elsewhere? We cannot give an adequate answer to that question, as we know only one form of life—that on Earth.

Our growing awareness and understanding about the origins of life on Earth, and possibly in many other similar or significantly dissimilar places in the Universe, and the likelihood of encountering extraterrestrial life, have raised a plethora of cultural, social, ethical, and religious questions. How are we to treat alien life-forms? How will they affect us? What if they are threatening or dangerous—or much more advanced that we are? Should we look for them? And how should we do that—what precautions should we take? What should we do when we finally discover alien life, or likely alien life—either directly or remotely? It is envisaged that microbial life will likely be more abundant in the Universe than intelligent life, and there is a range of well-known issues surrounding the possibility of extraterrestrial intelligent life, and how its discovery would impact us, as well as how we could or should communicate with it and interact with it.

There are also other, perhaps more profound and immediately relevant questions that astrobiology and the possibility of life elsewhere provoke. What does this tell us about who we are? Are we unique? Are we the only ones who know about the Universe? How does this knowledge and understanding affect our attitudes toward ourselves as individuals and as a society—and toward environments and life-forms, conscious or not, here on Earth and beyond Earth?

Before we can answer these and similar questions, which go beyond the usual scientific and technological concerns, we need to develop reliable criteria and a framework within which we can decide such issues. In doing that, it is essential to appreciate and acknowledge our own relationships with life here on Earth and with possibly existing life elsewhere. This naturally leads us to a sense of our deep, often (up to now at least) hidden, but essential connections and dependences on the characteristics of our Universe, our Galaxy, our Solar System, and our Earth and biosphere. This leads us to an expanded and much truer view of who we are and what our overlapping societies are—as products of cosmic and biological evolution, but also as knowers and (hopefully) responsible actors and agents in this ongoing drama. Added to that are a growing understanding and appreciation of the life-forms and environments here on Earth and elsewhere—and their value in themselves and within the environments of which they are a part. That includes their complexity and organization, their capabilities, and their individual and communal behaviors—but also and very importantly how they might evolve in the future. This strongly affects and modifies our appreciation of them and their relationships to us, as we see in our own increased sensitivity toward the other creatures and environments we share our planet with and toward those plants and animals from which we have descended by evolution.

What are our attitudes and views toward the world and the Universe in which we live? What should they be, what influences them or determines them? On what basis do we assign levels of value, goodness, respect, sacredness to entities we are connected with? We do so at least partially in light of our scientific knowledge and understanding, but also partially in terms of the intrinsic good of the *other*—not simply as a good for us. Only with these considerations in mind can we robustly attribute ethical status to other systems, life-forms, societies, and environments, and only then can we answer the ethical, cultural, and religious questions raised by astrobiology.

Astrobiology: State of the Field

For discussion of the ethics of astrobiology to be meaningful in more than a hypothetical way, there has to be some realistic prospect that these issues can be made concrete. That will happen only if and when we discover life on another world (or create it in Earth's labs). Astronomers are sanguine that the search is very well motivated; since the time of Copernicus the "principle of mediocrity" has been affirmed at every turn—there is nothing special about the physical nature or cosmic environment of our planet, our star, our Galaxy, or our region of the Universe. After thousands of years of speculation, there are finally signs that the search for life beyond Earth may soon succeed.

Astrobiology is focused on three basic questions: How does life begin and evolve? Does life exist elsewhere in the Universe? What is the future of life on Earth and beyond (Des Marais et al. 2008)? The NASA Astrobiology Roadmap identifies seven goals that expand on these three basic questions, with the life sciences being the driving theme. Researchers will endeavor to make biology and life detection a central component of future missions to planetary bodies in the Solar System, in the search for Earth-like extrasolar planets, or exoplanets, and in theories and models of the origin of the Universe. Other goals are to foster research to better understand life on Earth—its history, evolution, diversity, and limits—with a goal of increasing our chances of detecting life elsewhere. Life may turn out to be a natural and inevitable product of cosmic evolution.

The past 20 years have seen accelerating progress on a number of fronts (Impey 2010, 2011). Ironically, the era began with a false dawn: the claim that there were fossil microorganisms in the Martian meteorite ALH 84001 (McKay et al. 1996). Although that evidence was later not found to be compelling (Jakosky et al. 2007), it sowed the seed for a new NASA initiative in the study of "origins of life, planetary systems, stars, galaxies, or the Universe" (Smith 2004). In 1997, NASA established the NASA Astrobiology Institute (Dick 2007). Within three years of the AHL 84001 findings, the *Mars Global Surveyor* orbiter returned images that showed the strongest evidence that liquid water had once existed on the surface of Mars. The *Mars Odyssey* orbiter, launched in 2001, added dramatic evidence from hydrogen spectral measurements that extensive areas of water ice are below the surface of Mars. More recently, the Mars Exploration rovers confirmed that extensive quantities of liquid water had existed on the surface and in the subsurface of Mars in the past (Ehlmann et al. 2009). The fact that methane is escaping into the atmosphere from the subsurface of Mars (Mumma et al. 2009) adds to the sense that Mars has the ability to support subsurface microbial communities even today. NASA's *Curiosity* rover, currently in the early phase of its mission, is expected to affirm that Mars was habitable in the distant past and may be habitable in particular locations even today.

Besides significant results from the Mars orbiters and rovers, there have been other important breakthroughs in astrobiology-related research. Observations with the Hubble Space Telescope led to the discovery of disks of gas and dust around young stars that are thought to be the precursors of planetary systems. Observations from the *Galileo* spacecraft suggested the presence of a liquid ocean below Europa's icy surface. The *Cassini-Huygens* lander revealed that Titan has organic solvent lakes composed of ethane and methane, suggesting that this moon of Saturn may be an exciting laboratory for understanding prebiotic biochemistry. If life exists on Titan, its biochemical basis will be different from that of life on Earth. *Cassini* also made close passes by Enceladus, another moon of Saturn, and found water-rich geysers spewing organic compounds. This small world surprised everyone by having all the ingredients assumed essential for life: organic material, liquid water, and a local energy source (Matson et al. 2002). There may a dozen habitable "spots" in the Solar System.

Alongside exploration of the Solar System, progress has been made in understanding the only form of biology we know of. New molecular methods have unveiled a huge and diverse group of previously unknown microorganisms and viruses, referred to as "the unknown biosphere." It is estimated that there may be more than one billion species of microorganisms, of which less than 10,000 have been characterized (Ogunseitan 2004). The physiological diversity of these unknown microbes, and particularly of those from extreme environments, may greatly expand the environmental limits of carbon-based life. Another application of modern molecular tools is to control evolution, and even to construct organisms with the ability to grow under environmental conditions only found on other planetary bodies. This line of research could ascertain the actual environmental limits of carbon-based life, and it is a capability fraught with ethical issues.

The "explosion" in the number of known extrasolar planets is perhaps the most exciting phenomenon in all of science. Two decades ago, no planets were known beyond the Solar System, more than a few researchers had been burned by claims of detections that did not hold up, and many others had given up on the chase. When a planet with half the mass of Jupiter was found whipping around the star 51 Peg every four days, it was a stunning surprise (Mayor and Queloz 1995). We should, however, spare some surprise for the earlier discovery of planets around a pulsar, demonstrating that expectations are meant to be defied in astrobiology (Wolszczan and Frail 1992). Since 1995 the number of confirmed extrasolar planets has had a doubling time of 30 months, with a total of over 900 confirmed as of early June 2013. Alongside the growing numbers is the steady downward march of the detection limit from Jupiter mass to Neptune mass to Earth mass. Currently there are roughly a hundred extrasolar planets in the range of 2 to 10 Earth masses, colloquially referred to as "super-Earths," and over 50 extrasolar planets in the habitable zones of their parent stars—the range of distances within which water can be liquid on the surface of a terrestrial planet.

About 10 percent of Sun-like stars have planets, with indications that the true percentage may be much higher and that rocky terrestrial planets may outnumber gas giants (Marcy et al. 2005). NASA's *Kepler* mission has blown the lid off the search for low-mass planets. This modest telescope in a stable space environment can readily detect the 0.01 percent depth of eclipse caused by an Earth-like planet transiting a Sun-like star. The team announced over 1,200 candidates in early 2011, over 50 of which were in their habitable zones, among which 5 are probably less than twice the Earth's size (Borucki et al. 2011). By early 2013 the number of candidates had grown to over 2,700, over 350 of which are less than 1.25 times Earth's size (Batalha et al. 2013). It is just a matter of time before Earth-like planets are found in Earth-like orbits. Simple arguments suggest that habitable "real estate" around dwarf stars exceeds that around Sun-like stars, motivating new wide-field surveys for transits associated with stars much nearer to and brighter than *Kepler*'s faint targets. Extrasolar planet research is a burgeoning but still young field with many theoretical puzzles to solve before we can confidently project the number of habitable worlds (Baraffe et al. 2010).

The search for life in the Universe embodies a tension between the expectations sculpted by the nature of terrestrial biology and the sense that there may be some major surprises awaiting us in a Universe of roughly 10^{23} stars. In astrobiology, we have to draw as many inferences as possible from a single example of life—terrestrial life branching off from a Last Common Universal Ancestor—and we still have no general theory of living systems (Baross 2007). Yet life on Earth has been shaped by the geological and chemical context of this planet, and life elsewhere would naturally be subject to a different set of constraints. A microcosm of this tension is seen in the search for life on Mars. Astrobiologists set the framework for life detection on or near the surface of Mars using analog environments on the Earth (Doran et al. 2010), but from the *Viking* landers through to the *Mars Science Laboratory*, it is very difficult to gather evidence for, or decisive evidence against, terrestrial biology (Klein and Levin 1976). Experiments like these are designed to detect organisms with nucleic acid information storage and familiar metabolic pathways. If life has ever existed or still exists on Mars, it may be at or beyond the envelope of conditions tolerated by terrestrial extremophiles (DasSarma 2006).

If the necessary and sufficient requirements for biology are water, organic material, and free energy, then those minimum requirements are likely to be met in diverse planet and moon settings in our Solar System and beyond. Extremophiles tell us that biology is possible with only episodic access to water or with water that veers to above the boiling point and below the freezing point. There are many sources of energy in astrophysics: chemical, geological, gravitational, and magnetic, to give an incomplete list. The potentially subtle signatures, or biomarkers, of these habitable environments will be very difficult to identify as we turn our attention to the Universe beyond our backyard (Kaltenegger et al. 2010).

Biomarkers are required to take the huge step forward from habitability to the first detection of life beyond Earth. That detection, which is keenly anticipated by all astrobiologists and by members of the general public with an interest in science, might come in the form of a shadow biosphere on our planet, or from trace fossils in a Mars rock, or from future diagnostics of targets in the outer Solar System, or from a spectral signature imprinted in the atmosphere of an extrasolar planet, or from success in the long campaign to detect signals from remote technological civilizations. Each of these possibilities implies a different type of evidence, which must be matched against uncertain criteria for the definition of success.

The search for extraterrestrial intelligence (SETI) vaults over all the uncertainties regarding the abundance and evolutionary paths of biology beyond the Earth and attempts to look for artificial signals from advanced life-forms (Tarter 2004). SETI has been met with over 50 years of the "great silence" as researchers detect only radio static from stars that might host habitable planets. The practitioners are not discouraged, however, because advances in computation and the detection of narrow bandwidth radio signals mean that the exploration of the domain where signals might be detected is in its infancy. The Allen Telescope Array promises to place SETI in the regime where the sensitivity to equivalent versions of our technology will span millions of Sun-like stars, and any continued non-detection of artificial signals will mean that technological civilizations like ours are rare (Welch et al. 2009). For many astronomers SETI is an essential "side bet" in astrobiology, worth the effort and the null results because we know so little about the outcomes of biological experiments and because the implications of making contact would give a concrete form to many of the ethical issues explored in this volume.

The Copernican Principle has been robust enough to bear our weight at every turn in the long history of astronomy. Our situation on a rocky planet that orbits a middleweight star on the outskirts of an unexceptional spiral galaxy appears not be unusual or unique. A conservative estimate might be that there are a billion habitable "spots"—terrestrial planets in conventionally defined habitable zones, plus moons of giant planets harboring liquid water—in the Milky Way alone. That number must be multiplied by 10^{11} for the number of "Petri dishes" in the observable cosmos. Do we imagine that they are all stillborn and inert? Or do we imagine that a significant fraction of them host biological experiments, either like or unlike the experiment that took place on Earth? That is the central question of astrobiology. The ethical and philosophical implications of the compelling scientific experiment to search for life beyond Earth are the motivations for this volume.

The Relevance of Each Contribution

With a backdrop of vitality in research on astrobiology, there is ample motivation to consider its ethical implications in terms of the philosophy described earlier. Each of the chapters contributes to this philosophy and vision, and to its practical application in a certain specific way. Each chapter presents the voice of a researcher who considers not only the concrete details of scientific observations and experiments but the broader implications of his or her work. Physical scientists expand into the territory of philosophy, and philosophers expand into the realm of physical science, to create connections across the disciplines and move closer to understanding and actions. Here we indicate very briefly those connections that provide fertile ground for ongoing discussion and the development of more robust connections over time.

The book begins with chapters that provide background on the fundamental philosophical and religious traditions, to create a foundation for discussion of ethics and astrobiology and to construct the platform on which readers can anchor their consideration of proposals promoted in the later chapters by Margaret Race, Christopher McKay, Woodruff Sullivan, Jill Tarter, and Ted Peters. In chapter 2, Carol Cleland and Elspeth Wilson provide a careful, detailed, and profound philosophical and scientific examination of what life is. Scientists (and philosophers) now realize that defining life is as much a philosophical question as a scientific question. These authors also consider the roots of our ethical and moral obligations toward nonhuman organisms, including their intrinsic moral status—active and passive. That moral status cannot be determined independently of chemical, biological, social, and environmental definitions of life. The network of relationships and connections it enjoys is part of that ethically relevant reality. In particular, Cleland and Wilson challenge our narrow, anthropocentric viewpoint on the moral status of nonhuman life, and provide practical considerations to expand our appreciation and respect for it.

This stance is reinforced and deepened by William Stoeger in chapter 3. Stoeger explores the various approaches to the foundations of ethical reasoning, which have deep roots in Western philosophy. He then argues for a meta-ethical approach to relate this Western approach to the radically relational character of physical, biological, and sociological reality. This approach takes very seriously our scientific understanding of the Universe, of nature, and of ourselves within our environment, and connects it with our appreciation of the goods that these constitute and the consequent values and meanings they embody and disclose. Stoeger explicates a fundamental basis for judging what is good, what is of value, and what we are required to do ("oughts"), which provides a framework for critical reflection on special ethical, social, and cultural issues raised by astrobiology. He discusses the interdependent character of everything that constitutes nature and the framework within which we can assign priorities to the different goods, values, and oughts we are trying to balance in our personal and communal lives. The choices we make to act or not act, to pursue a certain program or not, are made based not only on our scientifically based knowledge and understanding, but also on our informed judgments of actual and expected short-term and long-term consequences and of what is good or better to do, on the priorities we have established as individuals and societies, and on the system of values and meanings that guides our engagement with reality.

Historical philosophical and religious points of view provide a useful perspective from which to explore astrobiological questions. In chapter 4, Martinez Hewlett gives a fascinating sketch of the informed philosophical and theological speculation and controversies about other worlds populated by living conscious beings. This, of course, undergoes important stimulus with the rise of evolutionary theory, and the dominance of explanation exclusively by physical and chemical processes. But is that all there is to being human, and self-reflectively conscious? Ontological reductionism does not seem to work—either a matter/soul dualism or some quantum-based correlated relationships may be needed.

Cleland and Wilson, Stoeger, and Hewlett represent the Western philosophical traditions, which differ in profound ways from Eastern philosophical traditions. In chapter 5, Nishant Irudayadason complements this Western astroethical view with his discussion of Eastern philosophical and religious perspectives. He brings key Buddhist and Hindu insights to bear on the harmony of the Universe, or "dependent co-arising"—the view that everything originates from a variety of causes that are mutually dependent—and the deep interconnectedness of all that exists. Along with this examination,

Irudayadason develops the three levels of good and evil in Buddhist thought, and shows how these intuitions strongly move us toward a view of nature and of ourselves that takes us beyond anthropocentrism to a more universal ethical perspective. Despite these different worldviews, there is a strong consonance between Irudayadason's views and those of Cleland and Wilson, Stoeger, and the later chapters by Sullivan and Peters.

After presenting the philosophical foundations that inform human actions and considering their application to astrobiology, the contributions move to more concrete applications. In chapter 6, Mark Bedau and Mark Triant examine the social and ethical issues arising in the work being done to create artificial cells (protocells). These are very closely related to many of the issues connected with astrobiology—especially those connected with potential benefits and risks, and the unknown character and biological impact, of such life, its moral status, the ethics of manufacturing life from scratch, and the commodification of life. In their careful analysis, they consider the various principles that have been suggested for making decisions in the creation of artificial life and that, by extension, have been applied to astrobiological exploration. Bedau and Triant find many of these commonly used principles wanting. Some, such as utilitarianism, decision theory, and the precautionary principle, are helpful, but none of them are really adequate to making important decisions, which involve unforeseeable consequences ("decisions in the dark"). Bedau and Triant recommend proceeding with informed courage balanced by wisdom and caution.

To give the discussion of new frontiers concrete substance, in chapter 7 Margaret Race surveys efforts to address the new frontier in space exploration. She outlines what has already been accomplished by policy agreements, including planetary protection policies. She describes how these policies are being revised and subjected to further revision in light of a whole range of societal issues, including but not limited to the ethical. She also emphasizes the importance of bringing expertise and participation beyond that of the scientific community to bear on these issues. Environment, health, and safety have been well emphasized, but there is a whole range of other social, ethical, sociopsychological, and religious concerns, stemming from the impact of astrobiological research and its successes, that have yet to be addressed in detail.

In chapter 8, Christopher McKay champions an ethical stance that privileges life as the most important source of value. This would demand not only that we promote and protect the richness and diversity of life on Earth, but also that we seek out and support independent instances of life elsewhere, and expand life from Earth to other planets, where appropriate. Following these imperatives requires that all our extraterrestrial exploration be biologically reversible to protect other possible extraterrestrial life-forms. We should also begin to experiment in a careful, limited, and biologically reversible way to establish terrestrial life on other planets in our Solar System. Mars is the most proximate and particular goal of this approach. McKay proposes that, if no other life-forms are already present there, Mars should be endowed with a terrestrial biosphere (called "terraforming") when this becomes technologically feasible and affordable.

This assertive approach is in contrast with more "hands-off" approach suggested by Woodruff Sullivan in chapter 9, although both are founded on "first do no harm" values. Sullivan presents a focused, concrete, and compellingly argued set of ethical principles to guide decisions in astrobiological research and exploration. They are based on principles adopted in environmental ethics as put forward, for instance, by an early American conservationist, Aldo Leopold. Sullivan's suggestions are consonant with those put forward by Stoeger, Cleland and Wilson, and Woolf. All five contributors strongly support the intrinsic value of extraterrestrial life and environments. But Sullivan goes further, elaborating the principles into concrete policies. In particular, he suggests that the decision to set aside Antarctica as an internationally protected and accessible resource be extended into a "planetocentric ethical policy" to protect extraterrestrial environments. While agreeing with McKay's privileging of life, Sullivan is very critical of the idea of expanding terrestrial life to other planets, such as Mars, that are potentially without life.

Expanding on Race's discussion of interactions with extraterrestrials in chapter 10, Jill Tarter takes an in-depth look at the search for extraterrestrial intelligence. She places particular emphasis on the scientific, social, political, ethical, and even journalistic challenges that a suspected detection or contact would present. She reviews the protocols already in place for responsibly dealing with successive stages of such a likely detection as well as with the overall impact its confirmation would have on society. A great deal of careful thought and discussion has already been devoted to preparing for such an eventuality. Tarter summarizes these debates and ventures an assessment of the likely influence such a discovery would have.

In chapter 11, Ted Peters further expands the astroethical discussion to consider how we should properly and ethically engage extraterrestrial intelligent life (ETIL) itself, should we encounter it. He goes beyond protocols for confirming, publicizing, and properly communicating an apparent first contact, or how to deal with primitive nonintelligent ET life. His discussion, in addition, provides a mirror to take a critical look at ourselves and see how we deal with one another here on Earth—particularly how we relate to and treat "the other." Peters suggests the ethical principles needed for dealing with three different categories of ETIL: those who are less evolved than we are; those who have reached approximately the same level of evolution and capability; and those who are significantly superior to us. Peters's suggestions should be compared and contrasted with Cleland and Wilson's discussion of the ethical status of ET organisms. Peters bases his ethics in these cases on responsibility—how should we respond in our engagement with ETIL, taking into consideration care for the health and welfare of life on Earth, and that of ETIL. As Peters says, "The conditions and imperatives arising from the new situation will suggest forms and frameworks within which to formulate our moral responsibilities." This flexible principle applies to dealing with ET life in general.

This approach overlaps with and complements what other contributors suggest. Peters elaborates an ethics of responsibility, which relies both on accumulating scientific knowledge and understanding of ETIL and its context, but also on the recognition of certain values, goods, and obligations with respect to that "other life." As we read and reflect on the scenarios explored by all the contributors, we are forced to examine more critically—from a more detached, and perhaps more objective point of view—our own values, priorities, and ethical principles, and how we engage, care for, and treat the life with which we share the Earth—how we engage, care for, and treat one another.

In the course of reflecting carefully on the meaning and importance of education, and specifically on "scientific literacy" and the stimulus astrobiology promises to give to achieving that societal goal, Erika Offerdahl, in chapter 12, emphasizes the increasing importance for us all to relate our scientific understanding and perspectives to our "philosophical and spiritual identities," which include social, cultural, ethical, and religious components. Echoing other commentators, Offerdahl points out in concrete terms the profound impact astrobiological discoveries and understanding already have—and will continue to have—on our social, cultural, and spiritual lives. Thus, in building educational goals and programs, we must give students the opportunity to develop the background, skills, and viewpoints that will enable them to grow in their interest, understanding, and appreciation of the natural sciences, and to relate this understanding to the broader issues and aspects with which they are concerned as individuals and as communities. She stresses that instructional methods need to foster scientific ways of thinking within an interdisciplinary context, which necessarily includes disciplines beyond the natural sciences themselves. This will prepare succeeding generations to face the challenges and opportunities arising from all scientific advances.

In chapter 13, Neville Woolf argues from his immersion in the science of astrobiology that life's highest priority is to retain all those achievements and features that make survival possible—not just basic survival, but the flourishing and long-term survival and enhancement of cooperation and understanding at their highest levels. Long-term consequences become important indicators of what is acceptable and what has ethical priority. What has led to fruitful outcomes over long periods of time should be embraced and fostered, and what has not should be carefully avoided.

The book closes with an appendix. Rather than an essay with a single voice engaging with results of new and challenging science, this contribution is a conversation or a dialogue between Steven Benner and Neville Woolf about the topics central to the theme of this book. Benner and Woolf discuss a set of research topics, including biological chemistry and synthetic biology and their application to constructing life different from terrestrial life, the task of understanding life as we know it and how various nonbiological compounds might affect it, and work to reveal the possible range of life elsewhere in the Universe and the conditions for its evolution. Both authors are scientists involved in astrobiological research. The astrophysicist, Woolf, and the biologist, Benner, provide a glimpse into the vital role that discussions bring to research, including the ongoing assessment of risks and benefits, along with developing protocols for countering and minimizing the risks and reaping the benefits. This dialogue is a taste of what the contributors experienced in their workshop discussions and what scientists and philosophers will be examining and debating in the coming decades.

The contributions to this volume cover a broad range of expertise, interests, and viewpoints. Each contributor brings a different foundation to the question we posed to them: How is progress in astrobiology stimulating social, cultural, and ethical issues, and how should we respond? Their contemplative responses are the product of many hours of discussion and thought, but they are only the starting point for expanding our views of ourselves, society, and the ethical implications of astrobiology.

References

Baraffe, I., G. Chabrier, and T. Barman. 2010. "The Physical Properties of Extra-Solar Planets." *Reports on Progress in Physics* 73: 016901.

Baross, J. A. 2007. "Evolution: A Defining Feature of Life." In *Planets and Life: The Emerging Science of Astrobiology*. Edited by W. T. Sullivan III and J. A. Baross. Cambridge: Cambridge University Press.

Batalha, N. M., et al. 2013. "Planetary Candidates Observed by Kepler: III. Analysis of the First 16 Months of Data." *Astrophysical Journal Supplements*, 204:24.

Borucki, W. J., et al. 2011. "Characteristics of Planetary Candidates Observed by Kepler. II: Analysis of the First Four Months of Data." *Astrophysical Journal* 736: 19.

DasSarma, S. 2006. "Extreme Halophiles Are Models for Astrobiology." *Microbe* 1: 120–126.

Des Marais, D. J., et al. 2008. "The NASA Astrobiology Roadmap." *Astrobiology* 8: 715–730.

Dick, S. J. 2007. "From Exobiology to Astrobiology." In *Planets and Life: The Emerging Science of Astrobiology*. Edited by W. T. Sullivan III and J. A. Baross. Cambridge: Cambridge University Press.

Doran, P. T., W. B. Lyons, and D. M. McKnight, eds. 2010. *Life in Antarctic Deserts and Other Cold Dry Environments: Astrobiological Analogs*. Cambridge: Cambridge University Press.

Ehlmann, B. L., et al. 2009. "Identification of Hydrated Silicate Minerals on Mars using MRO-CRISM: Geologic Context near Nili Fossae and Implications for Aqueous Alteration." *Journal of Geophysical Research* 114: E00D08. doi:10.1029/2009JE003339.

Impey, C. D., ed. 2010. *Talking about Life: Astrobiology Conversations*. Cambridge: Cambridge University Press.

———. 2011. *The Living Cosmos: Our Search for Life in the Universe*. Cambridge: Cambridge University Press.

Jakosky, B. M., F. Westall, and A. Brack. 2007. "Mars." In *Planets and Life: The Emerging Science of Astrobiology*. Edited by W. T. Sullivan III and J. A. Baross. Cambridge: Cambridge University Press.

Kaltenegger, L., et al. 2010. "Deciphering Spectral Fingerprints of Habitable Planets." *Astrobiology* 10: 89–102.

Klein, H. P., and G. V. Levin. 1976. "The *Viking* Biological Investigation: Preliminary Results." *Science* 194: 99–105.

Marcy, G., R. P. Butler, D. Fischer, S. Vogt, J. T. Wright, C. G. Tinney, and H. R. A. Jones. 2005. "Observed Properties of Exoplanets: Masses, Orbits, and Metallicities." *Progress in Theoretical Physics Supplement* 158: 24–42.

Matson, D. L., L. J. Spilker, and J. P. Lebreton. 2002. "The *Cassini-Huygens* Mission to the Saturnian System." *Space Science Reviews* 104: 1–58.

Mayor, M., and D. Queloz. 1995. "A Jupiter-Mass Companion to a Solar-Type Star." *Nature* 378: 355–359.

McKay, C. P., and H. D. Smith. 2006. "Possibilities for Methanogenic Life in Liquid Methane on the Surface of Titan." *Icarus* 178: 274–276.

McKay, D. S., E. K. Gibson Jr., K. L. Thomsa-Keprta, H. Vali, C. S. Romanek, S. J. Clemett, X. X. D. F. Chillier, C. R. Maechling, R. N. Zare. 1996. "Search for

Past Life on Mars: Possible Relic Biogenic Activity in Martian Meteorite ALH84001." *Science* 323: 924–930.

Mumma, M. J., G. L. Villanueva, R. E. Novak, T. Hewagama, B. P. Bonev, M. A. DiSanti, A. M. Mandell, M. D. Smith. 2009. "Strong Release of Methane on Mars in Northern Summer 2003." *Science* 323: 1041–1045.

Ogunseitan, O. 2004. *Microbial Diversity: Function and Form in Prokaryotes.* New York: Wiley-Blackwell.

Smith, D. H. 2004. "Astrobiology in the United States: A Policy Perspective." In *Astrobiology: Future Perspectives.* Edited by P. Ehrenfreund et al. Dordrecht: Kluwer.

Tarter, J. C. 2004. "Astrobiology and SETI." *New Astronomy Reviews* 48: 1543–1549.

Welch, J., et al. 2009. "The Allen Telescope Array: The First Wide-Field, Panchromatic, Snapshot Radio Camera for Radio Astronomy and SETI." In *Advances in Radio Telescopes.* Edited by J. Baars, R. Thompson, and L. D'Addario. Special issue, *Proceedings of the IEEE* 97: 1438–1447.

Wolszczan, A., and D. Frail. 1992. "A Planetary System around the Millisecond Pulsar PSR 1257+12." *Nature* 355: 145–147.

CHAPTER TWO

Lessons from Earth

Toward an Ethics of Astrobiology

Carol E. Cleland
DEPARTMENT OF PHILOSOPHY,
CENTER FOR ASTROBIOLOGY,
UNIVERSITY OF COLORADO–BOULDER

Elspeth M. Wilson
POLITICAL SCIENCE DEPARTMENT,
UNIVERSITY OF PENNSYLVANIA

If we assume that morality in other species will look just like human morality, we are likely to conclude that they don't have morality, having blinded ourselves to this fascinating aspect of their behavior. Rather, we need to proceed with an open mind and view each species on its own terms.[1]

What are our ethical responsibilities toward truly alien forms of life? This is an extremely difficult question to answer. In the absence of concrete examples of extraterrestrial life to reference and reflect upon, how can ethicists even begin to consider how humans ought to act in relation to organisms that we know nothing about? Rather than setting out to resolve this quandary, the purpose of this chapter is to identify and investigate some of the problems that might arise in attempting to apply classical theories in Western moral philosophy to extraterrestrial life. Our approach differs from others in this volume by examining the connections between empirical scientific research and theories within secular moral philosophy concerning the scope of our ethical obligations to nonhuman organisms. Most broadly, our goal is to highlight both the challenges we face in applying our predominantly human-centered theories of morality to alien organisms and to suggest potentially fruitful avenues for future research and analysis in this area.

Theories of morality traditionally distinguish moral agents from moral patients (or subjects). Moral agents have ethical obligations toward moral patients. The ability to reason and self-awareness are necessary conditions for moral agency. Only intelligent and self-conscious entities that understand that they can make choices and that their actions can harm and benefit others may be held morally responsible for their behavior. Life seems to be a precondition for moral agency. Nonliving entities (such as rocks, raindrops, tornados, and earthquakes) are not thought to be morally responsible for what they do, in spite of the fact that they sometimes cause massive environmental devastation and even catastrophic loss of life, because they both lack the ability to reason and are unaware of what they do. Although nonliving things may be said to have various kinds of value (for example, for their beauty, monetary worth, or contribution to the sustenance of life), they are rarely said to possess *intrinsic* moral status. In most secular moral theories, being a living entity is a precondition for having moral status (that is, being either a moral agent or a moral patient).[2] Even intelligent robots—which represent a possible counterexample to this principle—are often characterized in science fiction as constituting an artificial form of "life," suggesting that people are reluctant to ascribe moral agency to nonliving physical systems. When a fictional robot acts in an intelligent and self-conscious manner, we often conclude that it counts as an honorary "living being" rather than merely a "machine."

Because questions of ethics and moral status are so deeply rooted in the concept of life, we would be in a better position to understand our ethical responsibilities toward forms of life differing significantly from our own if we could provide a scientifically compelling answer to the general question: what is life? For this reason we will begin with this query in the first section below, but the focus of this chapter is not on the nature of life per se. Instead, it is on identifying the sorts of living entities to which we, as moral agents, owe moral consideration. Our concern is with the issue of what characteristics qualify a living entity—wherever and whenever it may be found in the Universe—as a moral subject. At one time the only living entities thought to be worthy of moral consideration were human beings, and many humans (women, racial minorities, the mentally impaired, and so on) were not accorded equal moral status. In recent years there has been a lively debate over whether our concept of moral status should be expanded to include some nonhuman animals, such as primates, dolphins, whales, dogs, and parrots. These and other issues are discussed in the second section below.

We conclude in the third and last section with some reflections on our ethical responsibilities toward different forms of life, both here on Earth and elsewhere in the Universe. We argue that the current human concept of moral status is problematically anthropocentric and Earth-centric. Just as our concept of life is bounded by our experiences with one example of life (namely, familiar Earth life), so our concept of moral status is (even more) narrowly bounded by our experiences with members of a single species of Earth life, *Homo sapiens*. Efforts by animal rights advocates to extend our concept of moral status to nonhuman animals are heavily influenced by how closely animals resemble humans in appearance and behavior. The possibility of very different forms of life existing beyond our planet highlights these anthropocentric preconceptions about morality. For we can envision discovering sentient forms of life elsewhere in the Universe that do not closely resemble us in either appearance or behavior. Indeed, it is somewhat ironic that it is easier for many people to take seriously the moral status of hypothetical extraterrestrials, particularly when portrayed as "intelligent," than the moral status of our actual closest nonhuman biological relatives on Earth, which we not only encounter within our own ecosystem but which also frequently look and act similarly to us. Organisms evolving in environments very different from Earth are likely to be radically unlike humans (Ward 2005), underscoring a need to philosophically scrutinize even our most basic anthropocentric and Earth-centric assumptions about morality.[3]

What Is life?

In most theories of morality, life is either implicitly or explicitly a precondition for having moral status. But what is life? Philosophers have struggled with this question since at least the time of the ancient Greek philosopher Aristotle, who was the first to try to define it (Aristotle, *De Anima*, bk. 2, 412a13–416b30).[4] In recent years the question "What is life?" has taken on increasing scientific urgency. Molecular biologists who are investigating the origin of life or trying to synthesize life in the laboratory from basic chemical building blocks want to know at what stage an ensemble of nonliving molecules becomes a primitive living thing. The answer is not clear. Likewise, astrobiologists charged with designing remote or *in situ* robotic experiments for detecting extraterrestrial life on other planets and moons worry that we might not recognize life if it differs significantly from familiar life on Earth. Even computer scientists find themselves mired in this controversy about the nature of life as they speculate about whether a sophisticated digital simulation of life could be truly alive.

The most popular strategy for answering the question "What is life?" is to formulate a "definition" of life. Unfortunately, every definition proposed thus far faces robust counterexamples. In the early to mid twentieth century, metabolic definitions were popular. They played a central role in the design of the equipment for life detection utilized by the 1970s *Viking* missions to Mars, NASA's first dedicated search for extraterrestrial life. At the core of all metabolic definitions of life is the notion of an object with definite boundaries maintaining its structure by exchanging matter or energy with its environment. Although prima facie appealing, metabolic definitions are deeply flawed. In their most general form they are too broad, because they include certain nonliving things such as candle flames, which maintain their physical form by extracting energy from their environment. Less general versions, on the other hand, are too narrow because they place restrictions on the material substrate doing the work, and hence exclude plausible alternative chemistries for life. The *Viking* experiments were based upon a very restrictive metabolic definition involving biochemical pathways utilized by microbes that are familiar to us on Earth. Attempts to find a middle ground, a definition that is neither too broad nor too narrow, have been unsuccessful because clever people invariably come up with scientifically credible counterexamples. There are good evolutionary reasons for supposing that life originating and evolving under chemical and physical conditions very different from the Earth's would not use the same chemical pathways as our form of life, even supposing that it was carbon-based. Moreover, biochemists still cannot rule out the possibility of discovering life-forms based upon the element silicon instead of carbon (e.g., Schulze-Makuch and Irwin 2006; Benner et al. 2004). In short, metabolic definitions of life are invariably either too broad (encompassing some nonliving organisms in addition to living ones) or overly narrow (excluding possible alien organisms differing from familiar life on Earth), and attempts to find a middle ground continue to be unsuccessful.

Analogous difficulties plague Darwinian definitions of life, which in recent years have supplanted metabolic definitions in popularity among biologists and astrobiologists. A widely cited version of this approach defines life as "a self-sustained chemical system capable of undergoing Darwinian evolution" (for instance, Joyce 1994). Yet it is not at all obvious that Darwin's famous mechanism for evolution, natural selection, is the only viable biological possibility. In recent years biologists have discovered some intriguing new mechanisms of biological change (such as epigenetic mechanisms of inheritance and lateral gene transfer) that at least prima facie question the exceptionlessness attributed to natural selection by the standard Darwinian model; whether they actually do so remains a hotly debated question among biologists and philosophers of biology. (We hasten to add that these mechanisms are natural biological processes and hence do not lend any support whatsoever to creationist accounts of the separate origin of the species.)

Many other definitions of life have been proposed, including thermodynamic, biochemical, cybernetic, and autopoietic definitions (see Popa 2004, appendix B, for a remarkably complete list), but like those just discussed, they all face serious counterexamples. This is not an accident. As Cleland and Chyba (2000, 2007) argue and we discuss below, the idea that life can be defined at all is fundamentally misguided, resting upon logical confusions about the nature of definition and its capacity to answer fundamental questions about natural categories.

The Nature of Definition

Definitions specify the meanings of terms. There are numerous types of definitions, ranging from the often circular "lexical definitions" supplied by dictionaries—to eat is to consume food, to hurry is to move fast, and so on—to "ostensive definitions" in which the referent of a term is specified merely by pointing, as a parent might do when teaching a young child what the color blue is (Look up at the sky, it is blue, and so too is the color of your shirt). The most informative definitions specify meanings by analyzing concepts and supplying a noncircular synonym for the term being defined; henceforth we shall refer to them as "ideal definitions."

The well-worn example of an ideal definition from courses in informal logic is this: "Bachelor" means unmarried, human male. What this definition says is that anything to which the term "bachelor" applies is something to which the complex expression "unmarried human male" applies, and vice versa; the quotation marks indicate that we are talking about a word (the word "bachelor") as opposed to the things designated by the word (a certain subclass of men). The expression on the right specifies *necessary* and *sufficient* conditions for the *application* of the word "bachelor." By "necessary" and "sufficient" we mean that each of these properties (unmarried, human, and male) is individually required for something to be properly labeled a "bachelor" and taken together enough for something to be labeled a "bachelor." The complex expression on the right ("unmarried human male") thus provides a linguistic analysis of the concept corresponding to the term on the left ("bachelor"). In other words, ideal definitions are concerned with *language* and *concepts*. They specify the meaning of a word by analyzing the concept associated with it into a set of necessary and sufficient conditions for the application of the word.

Ideal definitions work fairly well for terms such as "bachelor," "fortnight," and "chair." The thing to notice, however, is that the referent of each of these terms is a category whose existence depends upon human interests and concerns. Whether something qualifies as a bachelor or a fortnight is a matter of linguistic convention. There is nothing that bachelors have in common beyond being unmarried, male, and human, and nothing that fortnights have in common beyond being a time period of two weeks. Similarly, what counts as a chair depends upon its functional features vis-à-vis the human body. If a normal human can sit upright on it, then an object may be classified as a chair. The functional character of chairs allows them to be made of many different substances (wood, metal, plastic, granite, ice, and so on) and to come in a wide variety of shapes (oblate spheres, planar surfaces, curved surfaces, and so forth). The upshot is that chairs have little in common beyond their functional utility to people. Like bachelors and fortnights, what counts as a chair depends solely upon human interests and concerns. It is for this reason that the features that humans use to recognize things such as chairs, bachelors, or fortnights can be used to define the corresponding terms with remarkable precision.

Of course this is not to deny that there are borderline cases—cases in which it is genuinely uncertain whether something satisfies an ideal definition. A good example is the question of whether a 10-year-old boy is a bachelor. (Although most male children are unmarried, few of us would apply the term "bachelor" to a child.) Similarly, one might wonder whether two weeks plus an hour or two still constitutes a fortnight or whether a rock with an angled depression really qualifies as a chair. Moreover, even if one resolves such cases by adding additional conditions (such as "adult" to the right-hand side of the definition of "bachelor"), other borderline cases will invariably arise in turn, such as the culturally and legally debatable adult status of 16-, 18-, or even 20-year-old boys. The point is: language is vague. The imprecision of language is brought forcefully home by the classic example of trying to distinguish a bald man from a man who is not bald by counting the number of hairs on his head. At what point does a man become bald? How much hair must he have before he qualifies as bald? The fact that we cannot specify a crisp boundary (a determinate number of hairs after which a man ceases to be bald, or vice versa) does not show that there is no difference between being bald and having a full head of hair. Indeed, most of us feel that we can recognize a bald man when we see him. However, this example illustrates an important limitation often encountered in attempts to come up with accurate definitions: Genuinely ideal definitions—definitions specifying both necessary and sufficient conditions for the application of terms—are hard to come by, because we regularly encounter the logical quandary of borderline cases. Ignoring the problem of borderline cases, though, we can often come up with fairly satisfactory approximations of who counts as a bald man or a bachelor. Unfortunately, approximations of this sort, based upon shared social understandings, are not satisfactory when defining a scientifically significant term such as "life." If the counterexamples to the various popular definitions of life rested only on borderline cases (such as viruses), there would be no problem with them and we could settle with approximations. As we shall see, however, this is not the case.

Certain categories of phenomena, such as temperature, water, and lightning, are natural in the sense that they would exist even if there had been no humans to think or talk about them.[5] Had intelligent life never evolved on Earth, the unique chemical substance that we call "water" would still exist on Earth and elsewhere in the Universe. Natural categories are not amenable to a definitional approach. As an illustration, compare the term "water" to that of "bachelor." People normally recognize water by its sensible properties such as being wet, odorless, tasteless, and thirst-quenching. But unlike the case of bachelor, these characteristics do not tell us what water *really* is. Water has a nature that is independent of human interests and concerns. This nature is what the liquid in ponds, wells, aquifers, oceans, and raindrops have in common, namely, being H_2O. There was a time when people did not know that water was H_2O. Aristotle, for instance, thought that water was simple and unanalyzable, one of four basic physical elements (earth, fire, water, and air) out of which everything else is constructed. But even in those benighted days, water was understood to be whatever substance produced the sensible features that people used to recognize it, and even today we ordinarily rely upon these same features to identify something as water. If the meaning of the word "water" were (like that of "bachelor") fully captured by the characteristics that humans use to recognize it, then we would not be able to make sense of the claim that Aristotle was talking about the same stuff as modern chemists but he was wrong about it: Water is a compound, not an element. But the truth is,

Aristotle *was* talking about water *and* he got it wrong; today we know something about water that he did not know.

Natural Kinds

Water provides a good example of what philosophers call a "natural kind." Other examples include temperature, sound, lightning, and birds. Natural kinds differ from "artificial (non-natural) kinds" insofar as nature, as opposed to human interests and concerns, determines their membership. In the case of artificial kinds (such as bachelors, fortnights, chairs, and garbage), the properties we use to recognize them (their "*criterial properties*") can be used to capture the meanings of the terms that designate them; for this is all that the items falling under an artificial kind term have in common. In the case of a natural kind, however, something can fully satisfy all the criterial properties that we use to recognize it and *still* fail to qualify as a thing of *that* kind (Putnam 1975).

A good example is provided by the seventeenth-century controversy over whether bats are birds. John Locke (1689, bk. 3, chap. 11, sec. 7), who identified the meanings of all general terms with their criterial properties, viewed the debate as merely verbal—a matter of how we choose to define the term "bird." But we now recognize that this disagreement represented more than a mere verbal dispute, because we know that the physiology and anatomy of bats is typical of mammals and not birds. The classification of bats as mammals rests upon careful scientific investigations, as opposed to human concepts based upon superficial appearances. Similarly, something might lack the criterial properties associated with a natural kind term and yet on further investigation be classified under the term anyway. Diseases caused by the same microbes sometimes manifest themselves in strikingly different symptoms (for instance, tuberculosis and the bubonic plague), and thus may initially be classified as different maladies. What makes them symptoms of the same disease is their common causal origin. It takes empirical investigation, however, to discover this; mere analysis of concepts (based upon symptoms) will not reveal that they all share a common microbial cause.

It is time we applied these considerations to questions of the sort with which we began our discussion in this section, namely, questions of the form "What is X?," where "X" is a placeholder for a general term such as "bachelor," "chair," "water," and "bird." When X is an artificial kind term (such as "bachelor"), the criterial properties used to define it provide a good answer because we are describing a category delimited by human interests and concerns. In response to the question "What is a bachelor?," it is hard to do better than reply, "An unmarried human male." Anyone who disagrees simply does not understand the meaning of the term "bachelor" or, alternatively, is unfairly exploiting vagueness (borderline cases). In contrast, the question "What is water?" is not adequately answered in terms of the criterial properties used to recognize water, namely, being the wet, odorless, tasteless liquid that quenches thirst. (We most commonly think of water as a liquid, but it can become a solid at low temperatures and a gas at high temperatures, and under extreme pressures and temperatures it can take on even stranger properties.) The correct answer is: H_2O. The fact that water is H_2O cannot be discovered by relying upon normal human perception (what water looks or feels like *to us*) to define it. Moreover, it is important to keep in mind that the term "water" cannot be said to *mean* H_2O, because people grasped the concept of water long before they understood its molecular composition. In other words, no amount of analysis of our everyday concept of water will reveal that water is H_2O. This means that one cannot adequately answer the question "What is water?" by means of a definition of "water." Instead, what is required is scientific investigation in the context of the right theoretical framework—to wit, modern chemistry (see Cleland 2006, 2012, for more on this topic). This applies to all questions of the form "What is X?" where X is a natural kind term, including, as we shall see in a moment, the question "What is life?" Other salient (nonliving) examples are temperature (which is recognized by tactile experiences of warmth but is really kinetic energy) and sound (which is recognized by auditory experiences of noise but is really a compression wave). In a nutshell, attempts to "define" natural kinds such as water, temperature, and sound are fundamentally misguided because the most that can be revealed by conceptual analysis of the concepts that we associate with them is the way in which they are manifested to the human mind. What we really want to know is something else, namely, their fundamental nature as natural phenomena.

If life is a natural kind (as seems more likely than not), the same considerations apply to it. Generalizing over and abstracting from the criterial properties used by humans to recognize familiar Earth life is unlikely to reveal the nature of life. What we need in order to provide a scientifically compelling answer to the question "What is life?" is not a definition (an analysis of human concepts of life) but rather a general, empirically well-grounded theory of living systems. Unfortunately, as we discuss below, contemporary scientists are in no position to formulate such a theory. Familiar life on Earth represents a single—and quite possibly unrepresentative—example of life.

Life on Earth

We know a great deal about the chemistry and biology of life on Earth. Molecular biology has taught us that familiar Earth life is based upon an exceedingly complex cooperative arrangement between two types of large, organic (carbon-containing) molecules—proteins and nucleic acids. Proteins supply the bulk of the structural material for building organismal bodies as well as the catalytic (enzymatic) material for powering and maintaining them. Nucleic acids store the hereditary information required for reproduction and also for synthesizing the enormous quantity and variety of protein required by an organism during its life span. The crucial process of coordinating these functions—of translating the hereditary material stored in nucleic acids into proteins for use in growth, maintenance, and repair—is handled by ribosomes, minuscule but intricately structured molecular devices found in large numbers in the cells of all known organisms on Earth. The similarities go even deeper, however. Proteins consist of long chains (polymers) of smaller organic molecules known as amino acids. Although more than a hundred amino acids are found in nature, all life on Earth (bacteria, jellyfish, mushrooms, redwood trees, sharks, crows, humans, and so on) constructs its proteins from the same approximately 20 amino acids. Protein chemists have synthesized alternative proteins from abiotic amino acids, and they have shown that they would be functional in the right organismal environment. Yet although it is well within the realm of scientific possibility, no known organism on Earth utilizes such alternative combinations of amino acids to synthesize proteins.

Like proteins, nucleic acids are long molecular polymers. Instead of amino acids, they are synthesized from nucleotides. A nucleotide is a complex molecular unit consisting of three subunits, a phosphate unit bonded to a sugar (ribose or deoxyribose) unit that, in turn, is attached to a base unit. Familiar Earth life employs two different kinds of nucleic acid, DNA and RNA. With the exception of some viruses that use RNA, DNA is the repository of hereditary information and RNA supervises the intricate process of translating this information into proteins for immediate use by the organism. The DNA molecule utilizes four different bases—adenine (A), thymine (T), guanine (G), and cytosine (C)—to encode hereditary information. These bases pair off in a complementary pattern: C pairs with G, and A pairs with T. As a consequence, each strand of DNA provides a template for reconstructing the other half. This arrangement provides an important source of redundancy in case of damage to one strand. Hereditary information is encoded on a single (the "coding") strand of DNA by sequences of three consecutive bases. Each triplet of bases or "codon" specifies a specific amino acid or, alternatively, the initiation or termination of the construction of a chain of amino acids constituting a protein. With some minor exceptions, the genetic code is universal for all known life on Earth, with the same triplet of bases coding for the same amino acid. The translation of hereditary material into proteins occurs in ribosomes, tiny molecular devices made up of both protein and RNA.

The four bases used by familiar life are not the only molecular possibilities for storing hereditary information on duplex DNA; at least eight viable alternative bases have been identified (Benner et al. 2004; Benner and Switzer 1999). Moreover, the genetic code seems somewhat arbitrary. There is little reason to suppose that most codons could not have been associated with different amino acids. Instead of triplets, life might have utilized a doublet or perhaps even a quadruplet or quintuplet coding scheme. A triplet coding scheme is admittedly the most efficient available for four bases and 20 amino acids. But what if life had utilized a different number of amino acids to build its proteins or a different number of bases to code for amino acids? There is little reason to think that life could not utilize a different number of amino aids to construct viable proteins. There are at least six mutually exclusive base pairs (including the four used by our form of life) that could be accommodated by double-stranded DNA. It is not at all obvious that a triplet coding scheme would be optimal for organisms utilizing significantly more or less than 20 amino acids to build their proteins or more than two pairs of bases to build their nucleic acids.

In sum, despite its astonishing morphological diversity, all known life on Earth is remarkably similar at the molecular level. Many of these similarities do not seem to be biologically necessary. Molecular biologists and biochemists have established that life could have been at least modestly different in some specific ways—for instance, it could have utilized different amino acids to build proteins or different bases to build nucleic acids—and continue to speculate about more exotic biochemical possibilities (see, for example, Benner et al. 2004). The best explanation for these remarkable, seemingly contingent, similarities, in the face of so many striking differences in form and structure, is that life as we know it on Earth today evolved from an ancient (almost certainly microbial) common ancestor. If this is so, then our current understanding of life as a natural phenomenon—which is based solely on our experience with life here on Earth—is myopically Earth-centric. One cannot safely generalize from a single example of life to all life in the Universe. Indeed, we have no idea how different life could be from familiar Earth life. For all we know, Earth life may be fairly representative or very unrepresentative of life in the Universe. Until we accrue a wider range of examples—such as life not sharing a common ancestor with familiar Earth life or originating and evolving in extraterrestrial environments—we will be in no position to speculate about the nature of life in general.

Viewed from this perspective, astrobiologists should eschew the use of definitions of life and search for life beyond Earth with an eye toward new and unanticipated possibilities (Cleland and Chyba 2007; Cleland 2006). For they are likely to increase their chances of detecting "weird" extraterrestrial life if their equipment and search strategies are less Earth-centric—more sensitive to the possibility of encountering life substantially different from that with which we are familiar. This of course presents us with a conundrum: How can one search for something that differs in unknown but fundamental ways from what humans have experienced thus far? A number of strategies have been suggested for circumventing this problem (see, for instance, Davies et al. 2009; Cleland 2007; Nealson 2001). The most important lesson for our purposes, however, is that the Earth-centric human concept of life provides an inadequate foundation for reasoning about the possibilities for life in general. We just do not know what features of contemporary Earth life are the product of chemical and physical contingencies present on the early Earth (but perhaps rare on other worlds containing life), as opposed to characteristics that are universal to all life in the Universe.

Foundational Issues for an Ethics of Astrobiology

This brings us to the central question of this chapter: Given the highly Earth-centric character of our current understanding of life, how can we even begin to address the question of our potential ethical responsibilities toward forms of life differing radically from our own? We believe that this is arguably the most central ethical quandary facing astrobiologists pursuing the discovery of extraterrestrial life, because, as we shall discuss, one cannot understand the moral status of an organism independently of its chemical composition, physical structure, native environment, and way of life. But before we can pursue this question, it is important to first establish a clear notion of the difference between science and ethics.

Science and Ethics

Broadly, science is concerned with the discovery of empirical "facts"—both general, about laws of nature, and specific, about particular occurrences. With respect to the subject of this chapter, the job of science is to identify essential characteristics of life qua natural phenomenon. Ethics, in contrast, is concerned with describing how things ought to be or how we ought to act. The job of ethics is to evaluate issues of right and wrong (or good and bad), directing our focus to normative questions of value. In this regard science and ethics appear to operate in utterly different theoretical domains. As David Hume (1978/1738) famously argued, one cannot infer ethical judgments of *ought* from premises concerning only what *is*, because facts alone do not logically warrant evaluative ethical conclusions. We could hypothetically know all the empirical facts about a given situation, but this information alone would not tell us how we ought to act in the absence of proscriptive normative principles or a system of value. Moore (1903) makes a similar point about the relationship between propositions about the natural world and evaluative conclusions about goodness, which he terms "the naturalistic fallacy." And Max Weber (2004/1917/1919) emphasizes the importance of a similar "fact/value distinction" when it comes to political leaders making informed decisions about public policy. In short, one cannot logically deduce evaluative ethical conclusions from purely factual premises about the world.

Obviously ethics and science are logically independent enterprises, but this does not mean that the bodies of knowledge they accrue are not connected in important ways. There is a noteworthy association between our biological knowledge about an organism and the ethical conclusions that we reach about its moral standing. Indeed, we often refer to the presence of certain biological traits when making ethical claims. A common theme in debates about whether or not a living thing is deserving of moral consideration is for scholars to point to characteristics that are thought to biologically distinguish humans from nonhumans, including factors such as brain anatomy, advanced cognitive function, intelligence, the ability to communicate through language, and the capacity to engage in organized social behavior. Though ethics transcends science insofar as it reaches beyond purely factual questions about "what is" and addresses what "ought to be," the ethical possibilities that we are willing to entertain when theorizing about morality in applied settings are unavoidably dependent upon what we scientifically know (or presume to know) about the world around us. As Marc Bekoff and Jessica Pierce (2009, 30) recently put it, "'Do animals have moral behavior?' is a question that is neither pure science nor pure philosophy, and really we have to address both aspects of the question at once."

The physical and behavioral characteristics of a living organism obviously delimit (albeit do not determine!) the scope of the possible ethical conclusions that we are willing to consider about it. Likewise, ethical theory has the potential to provide scientists with morally significant normative principles about how they (and other human beings) ought to act in relation to various organisms. The intersection of these two fields is called bioethics. Astrobiology presents a new frontier in bioethics. The discovery of truly strange forms of extraterrestrial life would compel us to revise our Earth-centric concept of life in significant ways, which in turn would challenge our largely anthropocentric (and hence also Earth-centric) models of moral status. In this manner, purely factual questions about extraterrestrial life (its molecular composition, biochemical requirements, anatomy, physiology, and so on) and its interrelationship with its environment (for example, its sources of material and energy available, physical and chemical liabilities, intraspecies behavior, and relations to other life) are deeply intertwined with normative questions about how we "ought" to treat it.

Moral Status

What characteristics, then, must a living thing possess to qualify for moral status? As we mentioned earlier, moral status is commonly divided into two categories, moral agency and moral patienthood; we use the term "moral patient" interchangeably with "moral subject." Here we are primarily concerned with the concept of *moral patient*. Although all moral agents are de facto moral patients, some moral patients (such as young children and the mentally ill) are not full-fledged moral agents. The label "moral patient" identifies those beings toward whom we ought to act responsibly. Even though moral patients themselves are not always held fully responsible for their actions, we typically believe that they should have a certain level of moral standing because they are capable of being benefited or harmed by us. Because they can be wronged in a morally relevant way, moral patients have an ethical claim that applies to moral agents (such as ourselves), who presumably have the capacity to recognize and act upon that claim. As a result, how people treat them may properly be evaluated as right or wrong.

The agent/patient distinction is particularly important when considering the moral status of nonhuman animals. Although moral agents might have secondary ethical responsibilities to nonliving entities by virtue of their value to organisms holding moral status—consider, for example, natural features such as the ozone layer, historically significant objects like the Rosetta stone and the Egyptian pyramids, great works of art such the

Mona Lisa and the Sistine Chapel, and iconographic religious and national symbols like the American flag, a statue of the Virgin Mary, or the Star of David—these inanimate objects are generally not viewed as holding intrinsic moral status outside their relationship to people or other beings identified as moral subjects. Sometimes nonhuman animals are also treated as *secondary* moral subjects (much in the same way as the previously listed inanimate objects) in virtue of their value to humans. For instance, we might determine that we have indirect moral obligations to animals, based upon our own aesthetic, scientific, emotive, or ecological interests (see, for example, Baxter 1974; Norton 1988). Prominent examples of these human-centered views in environmentalism include: treating Lassie the dog with kindness only because she is dear to the boy next door; defending members of an endangered species like polar bears because biodiversity makes the world a more beautiful and interesting place; eliminating "invasive species" in places like Hawaii in the name of preserving a more pristine (or "natural") ecosystem for future generations to enjoy; and protecting top predators such as sharks and cougars because we value the role they play in maintaining a delicate ecological balance in their environment. In each of these instances, we label our ethical duties to the animals in question as indirect because the moral consideration we owe them is justified in reference to another goal. The animals are merely a means to another end, as opposed to an end in themselves.

Sometimes we speak of nonhuman animals as *primary* moral subjects having intrinsic moral value. For instance, although some environmentalists view animals as warranting only secondary moral concern, others treat them as primary moral subjects. They emphasize that the welfare and happiness of all animals (both humans and nonhumans alike) depend upon environmental factors, including the condition of the ozone layer, the protection of our rainforests and other ecosystems, the deleterious effects of pollution, and the growing impact of global warming and so forth (see, for example, Gruen 1993).

Our increasing willingness to entertain the possibility that some nonhuman animals have intrinsic or primary moral status is a relatively new phenomenon. Until recently, most theorizing (both theological and secular) about the scope of moral status was almost exclusively limited to human beings—and even then, as history ignominiously illustrates, many human beings have in practice been excluded from full and equal ethical standing on the basis of such morally arbitrary categories as race, sex, ethnicity, culture, sexual orientation, and religious affiliation.[6] Although we now widely recognize that such distinctions among humans are morally wrong, at least in theory, few of us would dispute the assertion that numerous forms of prejudice and discrimination (sexism, racism, homophobia, and so forth) remain serious practical political problems throughout the world, both at home and abroad, today. Indeed, our species (the members of *Homo sapiens*) has not yet lived up to the relatively recent international "human rights" goal of treating all of our fellow human beings with universal respect and dignity, a position that was famously articulated by the United Nations (1948) in the aftermath of the horrors of World War II in "The Universal Declaration on Human Rights."[7] The tendency of our own species to form "in-groups" and "out-groups" and to politically marginalize and exclude others (according to distinctions between "us" and "them") raises serious pragmatic concerns regarding how we are likely to interact with truly alien beings from other planets (see, for instance, Tajfel 1969).[8] Moreover, most people continue to think that there is an important ethical distinction between humans and all other animals. It follows that, here on Earth, we have yet to live up to the ideal of universal human rights or equal dignity among the members of our own species, let alone extend the boundaries of moral community to encompass a sense of respect and responsibility toward nonhuman animals. Indeed, given how common it is in discussions of moral status is to determine whether a living being is intrinsically deserving of moral consideration by pointing to characteristics that are thought to distinguish humans from nonhumans, this collapsing of morality into a predominantly "human" enterprise reveals just how deeply embedded the tendency toward anthropocentrism is in our traditional Western approach to morality. Even within the animal rights movement, nonhuman animals are accorded moral status as ethical subjects on the basis of characteristics that they have in common with people, not vice versa. Although the new discourse on "animal rights" is a noteworthy effort to move away from an anthropocentric view of morality—by expanding the circle of our ethical concern to encompass nonhumans—the primary "point of reference" in these new moral debates remains a conspicuously human norm. To illustrate, four basic categories of characteristics commonly come up as morally significant in Western secular and theological theory.[9] First, there is the factor of *intelligence*. Both ethicists and scientists often cite features such as rationality, memory, brain anatomy, organized cognitive function, the ability to reason, creative problem solving, abstract thought, the creation and manipulation of tools, and the capability to communicate as morally significant characteristics. Second, there is *sentience*. Sentience refers to the ability of an organism to experience suffering (Singer 1975). Typically a sentient animal is capable of feeling pleasure and pain, aware of its surroundings, and conscious of what happens to it. There are obviously many different degrees of sentience, and our own species *Homo sapiens* is almost always treated as the benchmark against which all other animal species are measured and classified. There is also *social behavior.* As a social species, human beings are inclined to regard features such as playfulness, expressions of emotion, communal lifestyles, group loyalty, acquisition of learned skills, complex power structures, and lasting family ties as morally significant. And we are particularly impressed when we see evidence of ethical social behavior within species other than our own, including cooperation, altruistic tendencies, filial and fraternal displays of affection, lasting friendships, long-term monogamy, the sharing of scarce goods, and demonstrations of what appears to be a sense of empathy, compassion, fairness, reciprocity, and even justice (Bekoff and Pierce 2009; Bekoff 2007). Finally, *the possession of an immortal rational soul* often comes up in many Western theological discussions about moral status, ranging from the theorizing of early Catholic theologians (such as St. Augustine and Thomas Aquinas) to many contemporary religious scholars (see, for example, Linzey 1994 for a theological critique of anthropocentrism).[10]

As this list of characteristics illustrates, when we think about ethics, regardless of whether we approach the topic from secular or theological angles, we tend to do so in anthropocentric terms. Most theological theories bestow upon human beings a special moral status in the grand scheme of things, although the exact significance of this status varies throughout different religious traditions. In the Judeo-Christian-Muslim tradition, having a "rational soul" or "divine spark" often plays the central role in determining moral status. And many of the most famous theologians have historically rejected the moral status of even our closest nonhuman relatives (chimpanzees, orangutans, and gorillas) on the grounds that they presumably do not possess a "rational soul" and unlike Adam and Eve were neither made in the image of God nor conferred "dominion" over nonhuman animals by God (see, for example, Genesis 1:26–29, 2:19, 9:3–6).[11] Such a view poses a serious problem for making sense of our ethical responsibilities toward extraterrestrial life, in addition to nonhuman animals here on Earth. Indeed, even if we are willing to expand the notion of a (rational) soul to encompass certain nonhuman organisms, it is unclear how we might objectively decide whether the members of a certain species possess a parallel notion of divinity, religious faith, belief in spiritual salvation, or relationship to God. Do nonhuman animals (whether terrestrial or extraterrestrial) believe in God or in a spiritual afterlife? How can we determine if they possess rational souls? And should evidence of their religiosity (or lack thereof) really shape whether or not we have moral obligations toward other species? In his studies of the circulatory system, the great sixteenth-century philosopher-mathematician-scientist René Descartes dissected living dogs and horses without anesthesia. He dismissed their screams of agony as analogous to the creaks and groans of damaged mechanical devices on the grounds that they lacked souls (Descartes 2003/1629). According to the Cartesian position, nonhuman animals are soulless automata (or merely "things"). Of course, there is ongoing disagreement among Western theologians on this issue, and Eastern religious beliefs on the subject often differ radically from Western, but theological theories about moral status tend to be highly anthropocentric because they emphasize a unique spiritual relationship between humanity and divinity (see, for example, Linzey 1994, particularly 3–27).

There is admittedly a long and diverse historical tradition of scholars, both religious and secular, postulating about whether or not extraterrestrials exist and what this might mean for humanity (for example, see Crowe 2008). Steven Dick, both an astronomer and a former chief historian at NASA, argues that "the anthropocentricity of our current conceptions of religion and theology . . . suggests that they should be expanded beyond their parochial terrestrial bounds" and that theologians are progressively making inroads toward meeting this challenge (Dick 1998, 253). Nonetheless, the central question with which Dick both begins and ends his popular book *Life on Other Worlds: The 20th Century Extraterrestrial Life Debate* is very different from the question we grapple with in this chapter. Dick asks his readers to consider "the meaning of extraterrestrial life for humanity" and how the idea of alien life has influenced "the perception of our place in the universe." We agree that questions such as these are important and provocative, yet it is also important to recognize that this manner of framing of the "extraterrestrial life debate" directs our focus *inward* at humanity (asking what "they" might mean for "us"). In contrast, the challenge we raise here is to look *outward* by examining the moral responsibilities we (human beings) might have toward truly alien forms of life.

We believe that secular approaches have a significant advantage over theological ones when theorizing about the moral status of nonhuman organisms. Whereas religions tend to appeal to theological principles that people who are not religious or who are members of other faiths might not be able to reconcile with their worldview, secular ethics attempts to formulate principles and criteria for morality that do not depend upon any particular religious doctrine or conception of divinity. As a consequence of the brute *fact of religious pluralism* (that people are passionate about divergent religious and nonreligious worldviews), the appeal of secular ethics is potentially broader than that of theology. But despite this inherent advantage, secular moral theory, as it currently stands, also suffers from an excessively human-centered concept of moral status.

Secular Ethics

There is a long secular tradition in Western moral philosophy. Although it would be impossible to do justice to the full range of these works within the confines of this chapter, it is worth noting that utilitarianism and duty ethics (deontology) are perhaps the most influential and widely cited contemporary examples of secular approaches to the issue of moral status. While differing in methodological orientation, both focus on distinctly "human" characteristics when ascribing moral status.

According to Immanuel Kant's classic version of duty ethics, the status of moral agency is conferred by *personhood*. Kant argues that all rational beings are "persons," and all persons are "moral agents" having the capacity to determine moral imperatives and consciously will their own ends. All other animals (or nonpersons) he explicitly labels as mere "things" (Kant 1785). For instance, in his *Lectures on Anthropology* (1798), Kant explicitly states,

> The fact that the human being can have the representation "I" raises him infinitely above all other beings on Earth. By this he is a person . . . that is, a being altogether different in rank and dignity from *things*, such as *irrational animals*, with which one may deal and dispose at one's discretion [italics added] (Kant [1798] 7, 127).

This is an interesting passage for several reasons. First, human beings are presented as the prototypes for moral agency (personhood). Second, Kant explicitly excludes nonhuman animals on Earth from the category of persons by denying that they share our capacity for rationality. Finally, by associating personhood broadly with the capacity for rationality, Kant leaves open the possibility of an alien (exhibiting human-like rationality) qualifying as a "person." Kant's theory of moral status, however, would be difficult to apply to extraterrestrials very different from us. On the one hand, we can imagine a rational, self-conscious hive-like alien (analogous to the "Borg" on *Star Trek*) that is comprised of numerous individual creatures resembling humans but lacking an independent self-consciousness or sense of "I." On the other hand, it might be extremely difficult to recognize forms of rationality in organisms whose physical needs and lifestyles differ markedly from our own. Can we not conceive of creatures that are so alien that what seems to be entirely rational behavior from their perspective might seem irrational to us?

Writing on the issue of moral standing during the eighteenth century, Jeremy Bentham, the founder of utilitarian philosophy, presciently worried that humans might someday conclude that it is wrong to reduce nonhuman animals to "things" rather than viewing them as fellow living "beings" that are capable of suffering. The passage in which he famously makes this point by emphasizing that animals feel pleasure and pain like we do is worth quoting at length, for it offers important fuel for reflection:

> Other animals, which, on account of their interests having been neglected by the insensibility of the ancient jurists, stand degraded into the class of *things*. [original emphasis] . . . The day has been, I grieve it to say in many places it is not yet past, in which the greater part of the species, under the denomination of slaves, have been treated . . . upon the same footing as . . . animals are still. The day may come, when the rest of the animal creation may acquire those rights which never could have been withholden from them but by the hand of tyranny. The French have already discovered that the blackness of skin is no reason why a human being should be abandoned without redress to the caprice of a tormentor. It may come one day to be recognized, that the number of legs, the villosity of the skin, or the termination of the sacrum, are reasons equally insufficient for abandoning a sensitive being to the same fate. What else is it that should trace the insuperable line? Is it the faculty of reason, or perhaps, the faculty for discourse? . . . the question is not, Can they reason? nor, Can they talk? but, Can they suffer? (Bentham 1978/1781, ch. 17 footnote)

Until recently, nonhumans have generally been excluded from our moral radar. Bentham provocatively questions the basis for this oversight by focusing exclusively on a trait that animals share with human beings: the ability to suffer. His utilitarian logic appears to be that if animals are *similar* to us in this one crucial respect, then they might also warrant moral consideration. Bentham's view was extremely progressive during the eighteenth century, but it is important to note that he focuses only on the similarities between nonhuman animals and us, and consequently does not move significantly beyond a human-centered view of moral status.

This shortcoming occurs in other versions of utilitarianism. Another early utilitarian philosopher, John Stuart Mill, maintained (contra Bentham) that there are different classes of happiness, and ranked human intellectual pursuits at the top of his list of pleasures. Mill (2002/1863) famously asserted that it would be better to be a dissatisfied human than a satisfied pig.

Variations of both duty ethics and utilitarianism play central roles in contemporary discussions about animal rights. The animal rights movement represents a concerted effort to expand the concept of moral status beyond our own species, *Homo sapiens*, and include nonhuman animals in discussions of morality. It thus provides a useful reference point for thinking about how to approach ethical theorizing about extraterrestrial life. There are many philosophical arguments that support the moral status of animals, but the two most influential today are Peter Singer's utilitarian defense of animal liberation and Tom Regan's deontological defense of animal rights.

Singer, the best-known contemporary utilitarian philosopher, argues that it is morally unjustifiable to exclude from moral consideration nonhumans who have a capacity to suffer (see, for example, Singer 1975). All animals conscious of pleasure and pain can suffer and thus ought to be accorded moral status, according to him. Singer provocatively argues that giving moral preference to our own species is a morally reproachable type of "speciesism." (Here it is worth noting that Singer's argument closely parallels our own point about anthropocentrism.) He further maintains that every sentient organism ought to have its interests given equal moral consideration, but he draws an important distinction between the more commanding interests of "higher" self-aware organisms versus those of "lower" organisms that have the capacity to suffer but do not have a strong sense of themselves as independent entities with a past and a future. Singer presents us with an impressive effort to consciously move beyond traditional human-centered ethical approaches in Western philosophy. His brand of "preference utilitarianism" is open to the idea that nonhuman animals can suffer in ways that are specific to their own capacities and way of life, and his argument highlights a common but widespread unit of analysis (pleasure and pain) that we humans can generally understand and empathize with when we recognize its manifestation in others. Yet Singer's division between higher/lower organisms and focus on "self-awareness" and "sentience" leaves us with pressing practical concerns in the realm of *applied ethics*. For how can a person perceive and measure self-awareness (or even degrees of pleasure and pain) in an accurate manner in nonhuman animals that may express such capabilities and perceptions in radically different ways than we do? If people seek to apply this utilitarian approach to animals, whose physiology and behavior defy or challenge our (human and Earth-focused) expectations, then Singer's ethics (which is admittedly quite sweeping in scope) still fails to provide us with the practical guidance and safeguards we need to escape the paradox of anthropocentrism.

Taking a different approach via duty ethics, Regan (1983) defends the moral status of animals using a rights-based method. He maintains that all "subjects-of-a-life" have inherent value, a term he applies only to self-aware sentient organisms capable of having beliefs and desires, consciously experiencing emotions, maintaining a psychophysical identity over time, acting deliberately to achieve ends, and conceiving of future goals. Regan asserts that all mammals older than approximately one year of age qualify as subjects-of-a-life, so long as they are mentally normal. Under such conditions, he maintains, animals possess inherent value and ought to be accorded moral rights. Although Regan and Singer use different philosophical approaches, they agree that sentience serves as a minimum threshold for moral status and that nonhuman animals can possess (in varying degrees) the morally relevant characteristics that qualify humans as moral subjects. Both scholars make important inroads toward expanding our circle of concern to include a wide array of nonhuman animals, but there remains a lingering element of anthropocentrism in each theory.

In general, scholars defending animal rights tend to focus on (interspecies) commonalities between humans and other animals. Such characteristics might be sufficient for ethical inquiry about life on Earth, but they are not necessarily broad enough to encompass the ethical possibilities that might arise in the event of encountering truly strange life beyond Earth. Because we have a common evolutionary past, it is not surprising that animals on Earth (including humans) share many similar traits and behaviors. Indeed, when we consider the aforementioned characteristics commonly cited as morally relevant, we see that none of these features is incontestably unique to humans. The more closely an animal resembles us physically and behaviorally, the more likely we are to identify with it emotionally and recognize it as exhibiting qualities we identify as morally relevant (Warren 1997).

John Fisher (1987) argues that our ability to sympathize with another animal—which typically requires an emotional sense of common identification—plays a fundamental role in how we determine the recipients for our moral concern. Without sympathy, he suggests, our ability to apply human moral principles to nonhuman animals is extremely limited.

Even today our attributions of moral status to animals are heavily influenced by the degree to which they superficially resemble or behave like us. Our closest nonhuman biological relatives in the animal world tend to attract greater moral sympathy from us than other animals do. For instance, we readily speak about great apes using familial terms, referring to them as our closest animal "relatives" and primate "cousins." In Spain, apes have recently been accorded certain basic legal rights against "abuse, torture, and death." Mammals are invariably given higher moral status than "fishy" cephalopods (such as cuttlefish and octopuses), despite extensive empirical evidence that the latter not only feel pain and pleasure but also have reasoning and memory skills rivaling those of monkeys. Domesticated human companion animals (dogs, cats, parrots, and so on) are routinely granted greater degrees of moral concern from us than food animals (such as cattle and chickens) and pests (insects, rodents, and so forth), even though many animals labeled as "vermin," like mice and rats, display remarkable levels of intelligence, ingenuity, and complex social behavior. In short, we seem to evaluate the moral worth of other creatures in relation to our own characteristics.

Our thinking about nonhuman morality suffers from human essentialism, or what Marc Fellenz (2007) has recently termed "species solipsism." Of course, a certain degree of human-centeredness is unavoidable in any moral code constructed by us to address our own ethical concerns and guide our sociopolitical way of life, but we can (and should) attempt to question and overcome our most glaring anthropocentric tendencies when we extend moral theory to nonhumans. Singer and Regan take important steps in this direction. However, the inevitable role of human self-absorption in formulating and applying these theories still raises serious concerns about our ability (or willingness) to actively seek out and recognize morally relevant qualities and behavior in other species, particularly in those that differ from us in significant ways. Although there are many characteristics that humans and nonhuman animals appear to share—and these commonalities might indeed serve as the basis for cultivating cross-species moral sympathy—Mary Midgley (1983) points out that we do violence to the sometimes immense and marvelous differences between our species when we focus purely on our similarities and ignore differences. The disparities between humans and various nonhuman organisms might be just as morally relevant to how we ought to treat them as our similarities. (To wit, merely because it would be wrong to dump a human being in the middle of the ocean without a lifeboat, this does not mean that the same moral principle applies to our interactions with dolphins and octopuses, both of whom require an aquatic environment to survive.) If we fail to judge the needs of each species with an eye toward the ways in which they differ from as well as resemble us, then we are likely to end up doing unintentional and potentially catastrophic harm to them by applying a procrustean conception of moral status.

Science fiction provides one of the best illustrations of the extent to which our current concept of moral status depends upon human attributes. Alien forms of life are typically accorded moral status only if they can be construed as resembling human beings in appearance and/or behavior. The more they look and behave like us, regardless of their origins or physical makeup, the more tempted we are to ascribe moral status to them. The same applies to robots. In the film *WALL-E*, a popular Disney-Pixar cartoon starring a heroic robot, our mechanical protagonist wins over our sympathy and moral respect by displaying a series of extremely "human" qualities, including a delightful curiosity about the world around him, solving problems in a creative fashion, expressing empathy toward others, forming lasting friendships, engaging in altruistic deeds, and ultimately falling in love with a beautiful "female" robot.

The original *Star Trek* series offers another instructive illustration in the episode "The Devil in the Dark." An alien creature (the Horta) resembling a "rock monster" causes catastrophic devastation by killing people and destroying equipment in a mining colony on another planet. Initially the Horta appears to be an unthinking monster—an amoeba-like blob of rock—dissolving everything in its path. But when Mister Spock is able to communicate with it by performing a Vulcan mind meld, he discovers that this creature is not an evil monster but instead a tormented "mother" defending her young from being destroyed by the miners. This ability to communicate and identify with one another transforms the Horta into a moral subject to the humans, and transforms the humans into moral subjects to the Horta, and both parties are in turn able to reach a mutually beneficial agreement. Given the Horta's lack of physical resemblance to human beings, the ethical standing attributed to the Horta depends crucially upon our being able to interpret its behavior as humanlike. (Indeed, what could be more human than a mother going to great lengths to save the lives of her children from what she perceived to be hostile intruders?) Yet why should we assume that moral status closely tracks our attributes or way of life? What if the Horta had physical needs, interests, and concerns radically different from ours? What if it was so divergent from us that communication was impossible? What if its intentions were beyond our scope of comprehension, and not simply a familiar example of a mother concerned about the safety of her children? Given the profound morphological differences between it and us, combined with the presumption that it evolved separately from us far beneath the surface of a distant planet, this would seem more likely than not. But as Gene Coon, the author of this *Star Trek* episode, must have recognized, it would be difficult to persuade television viewers that the Horta is a moral subject under such circumstances. If an alien creature looks radically unlike us, then the clearest way we can experientially relate to it (and hence empathize with its plight) is by making it think and act in a manner that is familiar to us.

The greater the physical and behavioral differences between us and another organism, the less likely we are to view it as having moral status. This poses a serious problem for generalization. As in the case of life, we have but a single clear-cut example: life on Earth. And although most people readily agree that human beings are moral agents, we continue to debate and puzzle about the moral status of other animals. The hostility faced by the animal rights movement underscores the extent to which even small differences in appearance and behavior can undermine claims about moral status. Science fiction fantasies aside, the initial temptation will be to dismiss the moral status of any extraterrestrial organisms we encounter, because they are likely to differ greatly from us in physical appearance and way of life, reflecting their origin and evolution in chemical and physical environments very different from our own. One can speculate about possibilities: They might be solitary individuals who do not care for their young, or so social (like the hivelike Borg) that it is difficult, if not impossible, to separate the interests of the individual from those of the group as we find in the "hive" behavior of the "Borg" aliens on the television series *Star Trek: The Next Generation*. (The group, as a whole, as opposed to the individuals comprising it, might be the basic unit for moral analysis.[12]) They might possess a technology or civilization so different from ours that it is difficult for us to recognize it as such, even though it might be extremely "advanced." Yet even these wild speculations are based upon our experiences with Earth life. An extraterrestrial race might differ from familiar life in ways that we have not imagined and yet still qualify as moral subjects to which we owe ethical obligations—or moral agents, which owe ethical obligations to us.

In the thought-provoking novel *The Black Cloud*, Fred Hoyle considers the grave difficulties that *we* might face in trying to convince an extraterrestrial that is far more intellectually advanced than ourselves that we are worthy of moral consideration! Like the Horta, the black cloud's moral agency is revealed through its having a conceptual structure and language that ultimately allows us to communicate with it and discover that, rather than being a nonliving mass of particles moving in space, it is a living creature that is not only self-conscious and extremely rational but also capable of feeling pleasure and pain like us. These are fascinating hypothetical examples of creatures that are radically different from humans in appearance and material composition but nonetheless display remarkable cognitive and emotive commonalities with humans that allow us to "bridge" our physical differences and view one another as moral subjects. But what if they were even more different from us and possessed bizarre manners of thinking and interacting with the world? The more difficult question to answer is how we can evaluate the moral status of entities whose physical appearance, molecular composition, and way of life differ greatly from our own.

The Moral Status of Nonhuman Organisms

Prior to discovering complex alien life, it is just as difficult to anticipate the variety of forms that moral status could take as it is to formulate a general scientific theory of life. Nevertheless, even in the absence of specific examples, we can challenge and expand the boundaries of our concept of moral status by exploring cases of nonhuman Earth life displaying what (upon philosophical reflection) seem to be morally relevant characteristics in spite of the fact that we do not always view these animals as moral subjects. The species that differ the most from us in physical appearance, behavior, and way of life will, of course, provide the greatest challenges.

Animal rights activists have long recognized that many mammals—primates (such as chimpanzees, gorillas, orangutans, and baboons), social carnivores (including wolves, coyotes, hyenas, and lions), cetaceans (dolphins and whales), elephants, certain rodents (rats, mice, and prairie dogs), and some birds (parrots, geese, and crows)—display characteristics associated in humans with moral status (Bekoff and Pierce 2009; Bekoff 2007). The great apes (chimpanzees, gorillas, and orangutans) provide the most compelling cases (see, for instance, Goodall 1986, 2000). They are our closest biological relatives in the animal world. Like us, they are highly social. When faced with challenging physical and social situations, they exhibit sophisticated capacities for reasoning and problem solving, fashioning tools and even weapons, and engaging in complex social interactions, including altruistic, cooperative, and empathetic behavior. They recognize themselves in mirrors, strongly suggesting self-awareness, and exhibit behaviors associated in humans with morally relevant emotions, such as compassion, guilt, grief, pleasure, and joy. And although they lack the physical ability to speak as we do, they have been taught to linguistically communicate with us through sign language. It is difficult to argue that the great apes are not moral subjects, but this is hardly surprising, given how closely they resemble our species. The ethical significance of their moral status (that is, what we morally owe them) is beyond the scope of this chapter, but few would disagree with the proposition that nonhuman primates are more than mere "things" to be disposed with as we please. As sentient beings (to a greater or lesser extent), nonhuman primates should not be forced to endure suffering at the hands of humans without extremely persuasive justification.

Like apes and humans, elephants are also highly social. They have long memories and display sophisticated problem-solving skills, including using tools to overcome challenging situations. They cooperate with each other in achieving goals, and exhibit behavior analogous to compassion, joy, and grief in humans. In Thailand an elephant named Paya paints what appear to be self-portraits and interpretations of the environment around her. In India bands of young male elephants cooperate in raiding villages for rice beer, working together to identify where in a village the alcohol is stored and then finding ways to procure it. (Thus, now in addition to drunken humans, some rural areas in India also must confront the problem of drunken adolescent elephants.)

Even some rodents exhibit behavior suggestive of moral status. An intriguing example is prairie dogs. Prairie dogs are highly social animals. Like humans, they live in communities with others of their species that are not related biologically to them, and even "kiss" upon greeting. Prairie dogs possess an astonishingly large vocabulary of calls, including warning calls specific to particular kinds of predators (hawks, coyotes, humans, cats, badgers, weasels, and eagles) and descriptive calls for properties such as direction, size, shape, and color (Slobodchikoff et al. 2009). They incorporate descriptive information into their warning calls, constructing strings of yips, bark, and chirps having a syntactic structure reminiscent of sentences such as "Here comes a man in a red shirt." Other prairie dogs respond to these vocalizations in highly discriminating fashions. Prairie dogs also coin new sounds when faced with objects (for instance, a black oval) or animals (European ferret) that they have never before encountered. Biologist Con Slobodchikoff is convinced that they have a primitive language and are literally talking with each other about their environment and perhaps other things as well. The capacity to communicate complex information to others about environmental hazards and opportunities seems strongly suggestive of moral status. Not only does it illustrate cooperation and reciprocity, but it also hints at concern for the welfare of others.

Departing further from humans in appearance and way of life, birds and cetaceans (dolphins and whales) also exhibit remarkably sophisticated cognitive abilities, social skills, and other behaviors that are associated in humans with moral status. Wild geese form lifelong monogamous relationships and exhibit reactions suggestive of extreme grief at the loss of a mate (Lorenz 1979). Crows are notorious tricksters that can count, solve arithmetic problems, and fashion primitive tools (Emery and Clayton 2004). Alex, the late great African gray parrot, could talk, count, and recognize colors, shapes, and sizes, and exhibited a grasp of the highly abstract concept of zero (Pepperberg 1999). In one famous instance, he verbally begged his guardian, scientist Irene Pepperberg, not to leave him overnight at a veterinarian clinic when he was ill, expressing concern that she was punishing him for a misdeed and promising to be a good bird in the future.

Although more closely related to us biologically than birds, cetaceans differ from us radically in both appearance and way of life. Yet they too exhibit complex behaviors indicative of "human" moral status (Schusterman et al. 2007). Cetaceans have long memories, exhibit impressive problem-solving skills, including tool use, when faced with physical and social challenges, and display behavior strikingly reminiscent of human expressions of compassion, grief, and joy. Dolphins assist each other in times of need (such as birth, injury, and illness) and frolic with people off swimming beaches. In birthing lagoons off the western coast of Baja, gray whales appear to seek out humans, get to know them, and then introduce their young to them. Cetaceans also exhibit fascinating linguistic abilities, the range of which we are only beginning to grasp. Dolphins are able to carry out grammatically complex requests (for instance, "right basket, left, Frisbee, in") communicated by human hand and arm movements. Some whales transmit messages over very long distances (hundreds of miles) by producing a wide variety of intense, low-frequency sounds (Rothenberg 2008). These communications are difficult for us to study and interpret because they are transmitted over great distances in water, as opposed to air, and cetaceans also use sound waves (sonar) for perception. Despite our limitations in studying cetacean behavior, there is strong anatomical evidence that supports the increasingly persuasive hypothesis that these animals are able to engage in complex forms of communication. Studies of whale and dolphin brains reveal neural structures associated in humans with higher cognitive functions such as self-awareness, empathy, and linguistic expression (Marino 2004).

Following Descartes, or the Cartesian view that nonhuman animals are nothing more than automata, some scientists dismiss all nonhuman animal behavior (however sophisticated and reminiscent of our own behavior, it appears) as amounting only to reflexive responses to stimuli or mere unreflective mimicry of human behavior. This possibility cannot be completely ruled out. Indeed, as philosophers have long recognized, the same strategy of behavioral interpretation may be employed toward one's fellow humans. Can you really be sure that your friend is suffering (as you would suffer) when she moans after a bike accident? You cannot actually feel her pain, so how can you be sure? Perhaps she is just exhibiting behavior reminiscent of pain, but not truly suffering. Is it truly compassion that moved a young man to help a confused old woman across the street? Did he perform this act out of sympathy (à la Hume), or purely out of a Kantian desire to do the right thing for the inherent sake of doing one's moral duty, or perhaps his act was intentional but stemmed from the selfish desire to gain esteem as "a kind and admirable fellow" from those witnessing the scene? Or perhaps he is only reflexively responding to conditioning in civility provided by his parents. We simply cannot know the precise inner motivations of another animal, whether human or nonhuman. But aside from a global (Cartesian) skepticism about the inner mental life of others, it is becoming increasingly difficult to defend the claim that no nonhuman animal experiences (as opposed to merely exhibits) morally relevant mental states such as rationality, pain, pleasure, aggression, joy, compassion, empathy, grief, and forgiveness.

The animals thus far discussed are fairly closely related to us biologically. This makes it easier for us to recognize seemingly morally relevant similarities between their behavior and our own in spite of our differences. The reluctance of many human beings to acknowledge the strong possibility that nonhuman animals exhibiting such behaviors have intrinsic moral status bodes ill for the prospects of recognizing moral status in extraterrestrial forms of life. Extraterrestrials are even less likely to exhibit behaviors resembling our own, particularly if they lack a technological civilization.[13] The significance of an organism's behavior (whether it is rational or expresses morally relevant mental states such as pain, pleasure, compassion, guilt, anger) critically depends upon its native environment and nature (physical constitution, needs, interests, concerns, and so on). Extraterrestrials originating and evolving in chemical and physical environments very different from those on Earth are unlikely to resemble us closely in appearance, behavior, or way of life. Because (as we argued in the first section) there is no reliable scientific way of predicting the possible physical and behavior forms of extraterrestrial life, we agree with Steven Dick (1998) that science fiction ("the realm of the imagination") provides thought-provoking speculation in this regard. However, though the speculations of science fiction authors are entertaining, they are constrained by the same anthropocentric preconceptions about moral status as the rest of us have. To revisit a couple of our previous examples: It is not an accident that the physically alien Horta's moral status is salvaged in *Star Trek* through the discovery that it is an anguished mother defending its children, or that the Black Cloud's moral agency is secured by its cognitive similarity (albeit vast superiority) to us. There is an anthropocentric basis to almost all existing approaches to moral status, even when we focus our attention on nonhumans. In order to address the limitations of our traditional conception of moral status (and ideally expand it to encompass forms of life quite different from our own), we also need to seek out real (as opposed to simply inventing fictional) cases arising at the boundaries of our traditional human-centered moral categories and methods of ethical reasoning.

Fortunately there are some intriguing candidates on our own planet. On Earth, arguably some of the greatest challenges to our anthropocentric concepts of moral status are found between two groups of invertebrates, cephalopods and hymenopterans. Cephalopods include octopuses, which are typically short-lived and asocial and have virtually no contact with their young, yet they possess large brains, which do not closely resemble ours anatomically. They also exhibit remarkably intelligent behavior (Mäthger and Hanlon 2008; Hanlon and Messenger 1996). Octopuses solve complex mazes, unscrew lids on jars to get food, and have even been known to dismantle aquarium pumps. They exhibit curiosity, playfulness, and boredom, and occasionally engage in what seems to be deliberate deception—for example, sneaking out of a tank at night to steal fish from other tanks and shutting the lids behind them when they return; only their wet tracks across the floor (and disappearing fish) give them away!

Octopuses have not, however, been observed to exhibit some of the emotive behaviors that we most closely associate with moral status, such as compassion, empathy, grief, and guilt. Does this reflect an intrinsic lack of moral sensitivity on their part? or merely our own limitations in understanding the behavior of creatures so different from us? Perhaps our anthropocentric biases, conditioned by our highly social way of life, get in the way of our recognizing how asocial organisms might manifest morally relevant cognitive states. In this context it is important to keep in mind that highly intelligent, asocial organisms lacking these affiliative emotions might still be able to recognize the moral worth of social organisms. Analogously, social organisms like us may have ethical responsibilities toward self-aware asocial organisms in virtue of their capacities for feeling pain and pleasure.[14] In other words, it is important for us to acknowledge the possibility that we may have ethical responsibilities toward organisms differing greatly from us, and vice versa.

Cuttlefish also challenge our concept of moral status. Like octopuses, they have short lives and big brains and are good problem solvers; they have been trained by researchers to swim through mazes in laboratories. Unlike octopuses, however, they are social and appear to learn from each other. Recent research reveals that they employ a large variety of highly complex signals to communicate with each other and other animals (Langridge et al. 2007). But these signals are not acoustic. They consist of patterns of color and texture extending over the entire surface of their bodies. These patterns are diverse and complex, and may change rapidly over short periods of time. Like prairie dogs (and some nonhuman primates), cuttlefish emit specific signals for a wide variety of different kinds of predators, and they apply clever, species-specific tactics when dealing with them. Intriguingly, they use polarized light to communicate secret signals to fellow cuttlefish while remaining camouflaged to other animals (Mather 2004). It is hard for us to interpret these signals, however, because they are so very different from modes of communication with which we are familiar. To put it bluntly: These creatures communicate through mechanisms that are so entirely unlike our own linguistic mode of communication that it is difficult for us to decode what they are conveying to one another.

Finally, certain insects display behaviors that humans associate with moral status, but they do so on a collective level. Many insects belonging to the order Hymenoptera (including bees, ants, and wasps) exhibit a remarkable capacity to build and maintain elaborate hive societies whose survival depends upon mass cooperation. Honeybees are a particularly interesting example. The honeybee colony is one of the most complex societies found in nature, consisting of a reproducing queen, several hundred male drones, and as many as 50,000 infertile female workers in a single hive. Despite their impressive population size, most honeybee colonies function smoothly as an integrated whole and make many decisions collectively. In *The Wisdom of the Hive*, Thomas Seeley (1996) describes how bees communicate specific information to one another by performing different types of dances—including the shaking signal, tremble dance, and waggle dance—that contain precise descriptions to important locations (such as food and nesting spots) and even reflect a bee's enthusiasm about the site. For example, when a colony is searching for a place to relocate (after it splits off from its original hive due to overpopulation), the bees appear to select their new home based upon a set of shared procedures and rules. Seeley and Visscher (2003) demonstrate that this process of decision making is unusually effective and efficient. First, the swarm sends out several small "scout parties" to scope out various possible locations. After the scouts return, each party describes its favorite location through symbolic dancing. The entire swarm then narrows the choices by conducting more-detailed investigations. Finally the colony selects a nesting location through a quorum method, which can be loosely compared to voting behavior in humans. Miller (2007) refers to this type of behavior as "swarm intelligence." Although individual bees are readily willing to sacrifice their own lives to protect the collective interests of the colony, the hive as a whole serves as an extremely impressive decision-making entity.

Whether octopuses, cuttlefish, or honeybees have intrinsic moral status as agents or subjects is an open question. What is certain, however, is that they challenge our concept of moral status in ways that anticipate the kinds of difficulties we may someday face in judging the moral status of extraterrestrials. At present these cases provide the best empirical grist available for philosophical musings about truly alien forms of life. If there are objective moral principles, or at least consistent ethical standards that we wish to stand by in the event of encountering extraterrestrial life, then surely these are important examples of organisms on our own planet that will help us transcend our anthropocentric preconceptions and achieve a better understanding of the potential ethical responsibilities that we might have to organisms differing dramatically from ourselves. When confronted with cases such as these, we are brought face to face with the anthropocentric limitations of traditional Western moral categories and ethical theory. The irony is that these examples are not truly "alien." Despite our many differences, there are strong reasons to believe that we have more in common with animals from our own planet than we are likely to share with extraterrestrials, particularly given the fact that all species discussed in this chapter share a common evolutionary ancestry and ecosystem with us here on Earth.

Conclusion

This chapter has raised more questions than it has answered. Our purpose has been to highlight some of the deep problems scholars face in applying traditional ethical models from Western philosophy to the realm of astrobiology. Not only do we reflexively tend to think about life in a myopic and Earth-centric manner, but we also have a habit of theorizing about moral status, the basic unit for ethical analysis, in narrow reference to a particular species of Earth life, *Homo sapiens*. The anthropocentric character of our conception of morality ought to give those working in the fields of bioethics and astrobiology serious pause for concern. As we have discussed, it is impossible to anticipate the variety of ways in which complex, behaviorally sophisticated forms of extraterrestrial life—adapted to environments very different from those on Earth—might differ from human beings. They might fail to resemble us in morphology, chemical composition, social structure, intelligence, and behavior, making it difficult, if not impossible, to apply our traditional human-centric moral categories. For how an organism manifests even highly abstract characteristics (such as rationality, self-awareness, and the capacity to suffer) critically depends upon its needs and concerns. Organisms evolving under chemical and physical conditions very different from those on Earth are unlikely to have a way of life that parallels our own.

In order to overcome the narrowness of our current conception of moral status, scientists and ethicists need to work together, searching for and evaluating evidence of moral behavior in new and unexpected places. Because we have yet to encounter extraterrestrial life, the best place to begin such a quest is on our own planet. As discussed in the third section of this chapter, Earth contains species that do not resemble humans closely in either morphology or way of life but nonetheless act in ways that are suggestive of morally relevant behavior. It is important that we keep an open mind when confronted with such evidence so that we do not reflexively defer to our preconceived notions about moral status, which might automatically exclude these animals from moral consideration. Although moral philosophy has focused almost exclusively on correlating moral status in nonhuman animals with their apparent similarities to us, *differences* between species may be just as relevant to how we ought to treat them. For differences among organisms are not always morally arbitrary. If a species is inherently asocial yet surprisingly intelligent (like the octopus), uses a strange method of communication that is difficult for us to understand (like the cuttlefish), or assumes a social organization very different from our own (like honeybees), then these differences can and should influence how we evaluate the species and interact with its members. Consequently, when we encounter another species here on Earth or elsewhere, we ought to make it a point to consciously pause to scientifically observe and ethically contemplate its behavior in the ecosystem in which it lives before drawing hasty normative conclusions about its moral status or the nature of our ethical obligations to its members. Although currently hypothetical, these quandaries about morality contain potentially important public policy implications for the theory and practice of astrobiology.

Notes

1. Bekoff and Pierce 2009, 20.

2. In many religions, transcendental lifelike properties are attributed to various nonliving entities. For instance, immaterial angels are believed to have rational souls, and rocks may have immaterial spirits. Our concern here, however, is with secular theories of morality; other chapters in this book cover theological perspectives on morality. In secular theories, life (qua physical process) is a necessary condition for being a moral agent or a moral subject.

3. For the sake of clarity, it is useful to specify precisely what we mean by the term "anthropocentrism." Our use follows that of scholars engaged in contemporary animal rights debates, where it is used to label and critique human-centered approaches to the moral status of nonhuman animals. As we emphasize in this chapter, even when moral philosophers attempt to expand our circle of moral concern to encompass nonhuman animals, moral theory continues to reflexively use humanity as a theoretical point of reference or norm (e.g., human anatomy, perceptions, and behavior) for reasoning more generally about the moral standing of nonhumans, whether terrestrial or extraterrestrial.

4. Aristotle was not only the founder of the definitional approach to understanding life but also the first to systematize the logic of definition (Aristotle, *Topics*, bk. 1; *Posteriori Analytics*; and *Metaphysics*). In his writings on the nature of life, he sometimes seems to define life in terms of "self-nutrition" (which corresponds to the modern concept of metabolism, that is, the use of material and energy extracted from the environment for purposes of growth, self-maintenance, and repair) and sometimes in terms of "reproduction."

5. An extreme metaphysical doctrine, radical antirealism (which includes some versions of idealism) holds that nature itself is created by the human mind and has no existence independently of our own. Like the vast majority of scholars, we do not accept this position. A weaker antirealist position holds that many of the properties that we attribute to natural phenomena are picked out in relation to human sensory modalities, which are the product of our particular evolutionary history on a planet having certain chemical and physical contingencies, and hence do not necessarily reflect bona fide divisions in nature or alternatively the properties that are essential to these phenomena. This position is compatible with our view, because it accepts the claim that there are natural categories independently of human interests and concerns; it merely emphasizes the difficulty we face as humans in identifying them.

6. This statement is self-evident. In fact, we would be hard-pressed to identify any culture or nation that did not exhibit certain forms of sociopolitical marginalization, exclusion, or group hierarchy (either in the past or the present). Struggles over power involving group-level categorization and prejudice appear to be endemic—albeit to a

greater or lesser extent—in the human political world and are not limited to any particular society or civilization, no matter how "primitive" or "advanced." For instance, see Brown (2006) for an erudite discussion of the theoretical underpinnings of the role of the "civilization/barbarism" dichotomy in sociopolitical history, from ancient Rome to the contemporary age of globalization, and Smith (1997) for an in-depth discussion of the pervasive role of discrimination throughout American political history.

7. The preamble of *The Universal Declaration on Human Rights* emphasizes our shared heritage as part of a common human "family," stating: "Whereas recognition of the inherent dignity and of the equal and inalienable rights of all members of *the human family* is the foundation of freedom, justice and peace in the world, [and] Whereas disregard and contempt for human rights have resulted in barbarous acts which have outraged the conscience of mankind . . . THE GENERAL ASSEMBLY [of the United Nations] Proclaims this Universal Declaration of Human Rights as a common standard of achievement for all peoples and all nations" (italics added). A full copy of this important international document is accessible online on the UN official website: www.un.org/en/documents/udhr/.

8. The school of psychology called Social Identity Theory, originally developed by Henri Tajfel and his student John Turner, makes important inroads toward explaining the apparent ubiquity of various forms of group prejudice in human society (see, for example, Tajfel 1969; Tajfel and Turner 1979). A Holocaust survivor, Tajfel sought to explain the prevalence of prejudice among seemingly normal individuals (consider, for instance, the oft-cited specter of the "good Germans," who were complicit with the rise of Hitler and the Nazi party). In his groundbreaking 1969 article on this topic, "Cognitive Aspects of Prejudice," Tajfel concludes—based upon a series of psychological experiments investigating the role of group categorization—that the psychological roots of group prejudice appear to be part of "normal" human intergroup social behavior. His experiments suggest that people tend to divide themselves into groups of "us" versus "them" (or "in-groups" and "out-groups"), based upon all sorts of available criteria (including factors such as neighborhood, occupation, race, religion, class, sports teams, hobbies, and political party). Tajfel posits that these groups give us a sense of identity and belonging in society, and they help us understand our relational place in the world. When applied to prejudice, Tajfel's evidence suggests that an "in-group" will stereotype and discriminate against the "out-group" to enhance their own self-image and in-group rewards, even in instances in which the "social categorization" (or group) in question is randomly assigned and has no previous social significance outside the context of a particular experiment (see, for example, Tajfel et al. 1971).

9. Many of these characteristics are discussed in detail in *A Companion to Ethics*, edited by P. Singer (1993); see esp. 343–344.

10. St. Augustine offers an influential summary of this position, which emphasizes the importance of humankind possessing a *rational soul*. In his classic text *City of God*, Augustine writes, "Thus God made man in his own image, by creating for him a soul of such a kind that because of it he surpassed all living creatures on earth, in the sea, and in the sky, in virtue of reason and intelligence, for no other creature had a mind like that . . . a soul of the kind I have described" (503).

11. Andrew Linzey (1994), a professor of theology and Catholic animal rights scholar, attributes this position, within Christianity, in large part to the influence of the acclaimed medieval theologian St. Thomas Aquinas, "whose theology still dominates

much Catholic thinking" (12). Following both Aristotle and Augustine, Aquinas articulated the position that humans, unlike nonhuman animals, possess rational souls, whereas "dumb animals and plants are devoid of the life of reason . . . they are naturally enslaved and accommodated to the uses of others" (*Summa Theologica*, Aquinas (1989), 125. It is therefore "not wrong for man to make use of them, either by killing them or in any other way whatever" (*Summa Contra Gentiles*, Aquinas [1989], 56). The Thomist position on animals merges the Aristotelian idea of animal irrationality (versus human rationality) with scriptural interpretations of Genesis, suggesting that animals exist by divine providence to serve human ends and thus have no intrinsic moral status. During the mid nineteenth century, Pope Pius IX refused allow the opening of an office to protect animals from cruelty, citing the Thomist notion that it was "a theological error" to conclude "that Christians owed duties to animals" (Pope Pius IX quoted in Cobbe 1874, 207). It is important to emphasize that such a position is not limited to Catholic or Christian theology. For instance, Richard Foltz (2006) highlights similar forms of animal exclusion (from moral concern) in traditional Islamic thought. Today some revisionist religious scholars, such as Linzey and Foltz, are increasingly challenging theological forms of anthropocentrism, emphasizing that the issue remains open to divergent theological and scriptural interpretations. In Linzey's words, "There are good biblical as well as biological reasons for sensing a continuity of living agents but also for positing greater capacities in spiritual self-awareness in humans and mammals" (9).

12. Another interesting science fiction example of hive creatures can be found in Orson Scott Card's *Ender's Game*. In this award-winning book, our young protagonist Ender is tricked into destroying an entire race of aliens, referred to as "the Buggers," during what he believes to be a mock simulation of a space battle he is commanding in the form of a video game. After unknowingly orchestrating the xenocide of the Buggers, Ender discovers an unborn hive queen who communicates with him psychically. She explains to him that her species, "the Formic," did not realize that human beings were sentient creatures when they first attacked Earth, because people exhibited such "alien" characteristics. The war resulted from this misunderstanding. Although the last surviving member of her species, the Formic Queen serves as the physical embodiment of the centralized group mind of her race.

13. It is not at all clear that we would recognize the significance of a truly alien technology, which is what one would expect from organisms differing from us in profound ways, but a discussion of this issue is beyond the scope of this chapter.

14. This argument about the potential moral status of asocial octopuses seems consistent with utilitarianism, but it is important to emphasize that utilitarianism alone is not broad enough to address all issues regarding moral status. For instance, a creature might have no ability to feel pleasure or pain as such, but might still possess moral agency. The android Data in *Star Trek: The Next Generation* is an interesting example. Although we are told that he is unable to directly experience the sensations of pleasure and pain (indeed, he is explicitly programmed not to suffer emotionally or physically), he is portrayed as a self-conscious, highly intelligent, exceedingly rational, and morally responsible entity that frequently goes to great lengths to protect the lives of his shipmates. The point is that the ability to suffer or experience happiness, as we typically understand these concepts (in relation to human beings), in itself is not general enough to encompass all entities that might warrant moral consideration.

References

Aquinas. 1989. *Summa Contra Gentiles*, bk. 3, chap. 2. In *Animal Rights and Human Obligations*, edited by T. Regan and P. Singer. New York: Prentice Hall, 1989.

———. 1989. *Summa Theologica*, ques. 64, art. 1. In *Animals and Christianity: A Book of Readings*, edited by A. Linzey and T. Regan. New York: Crossroads.

Aristotle. 1941. *The Basic Works of Aristotle*. Edited by R. McKeon. New York: Random House.

Augustine. 1984. *Concerning the City of God against the Pagans*. Translated by H. Bettenson. London: Penguin Books.

Baxter, W. F. 1974. *People or Penguins: The Case for Optimal Pollution*. New York: Columbia University Press.

Bekoff, M. 2007. *The Emotional Lives of Animals*. Novato, CA: New World Library.

Bekoff, M., and J. Pierce. 2009. *Wild Justice: The Moral Lives of Animals*. Chicago: University of Chicago Press.

Bentham, J. 1982. *An Introduction to the Principles of Morals and Legislation*. Edited by J. H. Burns and H. L. A. Hart. London: Methuen. Originally published in 1781.

Benner, S. A., and C. Switzer. 1999. "Chance and Necessity in Bimolecular Chemistry: Is Life as We Know It Universal?" In *Simplicity and Complexity in Proteins and Nucleic Acids*, edited by H. Frauenfelder, J. Deisenhofer, and P. G. Wolynes. Berlin: Dahlem University Press.

Benner, S. A., A. L. Ricardo, and M. A. Carrigan. 2004. "Is There a Common Chemical Model for Life in the Universe?" *Current Opinion in Chemical Biology* 8: 672–689.

Brown, W. 2006. *Regulating Aversion: Tolerance in the Age of Identity and Empire*. Princeton: Princeton University Press.

Card, O. S. 1991. *Ender's Game*. New York: Tor Science Fiction.

Cleland, C. E. 2006. "Understanding the Nature of Life: A Matter of Definition or Theory?" In *Life as We Know It*, edited by J. Seckbach. Dordrecht: Springer.

———. 2007 "Epistemological Issues in the Study of Microbial Life: Alternative Terran Biospheres?" *Studies in History and Philosophy of Biological and Biomedical Sciences* 38: 847–861.

———. 2012. "Life without Definitions." *Synthese* 185: 125–144.

Cleland, C. E., and C. F. Chyba. 2000. "Defining 'Life'." *Origins of Life and Evolution of the Biosphere* 32: 387–393.

———. 2007. "Does 'Life' Have a Definition?" In *Planets and Life: The Emerging Science of Astrobiology*, edited by W. T. Sullivan and J. A. Baross. Cambridge: Cambridge University Press.

Cobbe, F. P. 1874. *The Hopes of the Human Race, Hereafter and Here*. London: Williams and Norgate.

Crowe, M. J., ed. 2008. *The Extraterrestrial Life Debate: Antiquity to 1915 (A Source Book)*. Notre Dame: University of Notre Dame Press.

Davies, P. C. W., S. A. Benner, C. E. Cleland, C. H. Lineweaver, C. P. McKay, and F. Wolfe-Simon. 2009. "Signatures of a Shadow Biosphere." *Astrobiology* 9: 241–249.

Descartes, R. 2003. *Treatise on Man*. Amherst, NY: Prometheus Books. Originally published 1629.

Dick, S. J. 1998. *Life on Other Worlds: The 20th-Century Extraterrestrial Life Debate.* Cambridge: Cambridge University Press.

Emery, N. J., and N. S. Clayton. 2004. "The Mentality of Crows: Convergent Evolution of Intelligence in Corvids and Apes." *Science* 306: 1903–1907.

Fellenz, M. R. 2007. *The Moral Menagerie: Philosophy and Animal Rights.* Chicago: University of Illinois Press.

Fisher, J. 1987. "Taking Sympathy Seriously." *Environmental Ethics* 9: 197–215.

Foltz, R. C. 2006. *Animals in Islamic Tradition and Muslim Cultures.* Oxford: Oneworld.

Goodall, J. 1986. *The Chimpanzees of Gombe.* Cambridge: Harvard University Press.

———. 2000. *In the Shadow of Man.* New York: Houghton Mifflin.

Gruen, L. 1993. "Animals." In *A Companion to Ethics,* edited by P. Singer. Oxford: Blackwell.

Hanlon, R. T., and J. B. Messenger. 1996. *Cephalopod Behavior.* Cambridge: Cambridge University Press.

Hoyle, F. 1957. *The Black Cloud.* New York: Signet/Harper.

Hume, D. 1978. *Treatise on Human Nature.* Edited by L. A. Selby-Bigge. Oxford: Clarendon Press. Originally published in 1738.

Joyce, J. 1994. "The RNA World: Life before DNA and Protein." In *Extraterrestrials: Where Are They?,* edited by B. Zukderman and M. Hart. Cambridge: Cambridge University Press.

Kant, I. 1798. *Lectures on Anthropology.* Berlin: Akademie-Textausgabe as translated and cited in Gruen, L., 2012. "The Moral Status of Animals." *The Stanford Encyclopedia of Philosophy* (Winter 2012 Edition), Edward N. Zalta (ed.) URL=http://plato.stanford.edu/archives/win2012/entries/moral-animal/.

———. 1991. *The Metaphysics of Morals.* In *Kant: Political Writings,* edited by H. S. Reiss, 131–176. Cambridge: Cambridge University Press. Originally published in 1785.

Langridge, K. V., M. Broom, D. Osorio. 2007. "Selective Signaling by Cuttlefish to Predators." *Current Biology* 17: R1044–R1045.

Linzey, A. 1994. *Animal Theology.* Urbana: University of Illinois Press.

Locke, J. 1689. *An Essay concerning Human Understanding.* Oxford: Oxford University Press.

Lorenz, K. 1979. *The Year of the Greylag Goose.* New York: Harcourt Brace Jovanovich.

Marino, L. 2004. "Cetacean Brain Evolution: Multiplication Generates Complexity." *International Journal of Comparative Psychology* 17:1–16.

Mather, J. A. 2004. "Cephalopod Skin Displays: From Concealment to Communication." In *Evolution of Communication Systems: A Comparative Approach,* edited by D. K. Oliver and U. Griebel, chap. 11. Cambridge, MA: MIT Press.

Mäthger, J. A., and R. T. Hanlon. 2008. "Cephalopod Consciousness: Behavior Evidence." *Biology Letters* 2: 494–496.

Midgley, M. 1983. *Animals and Why They Matter.* Harmondsworth, UK: Penguin Books.

Mill, J. S. 2002. *Utilitarianism.* Indianapolis: Hackett. Originally published 1863.

Miller, P. 2007. "Swarm Theory: Ants, Bees, and Birds Teach Us How to Cope with a Complex World." *National Geographic* 212: 126–147.

Moore, G. E. 1903. *Principia Ethica.* Cambridge: Cambridge University Press.

Nealson, K. H. 2001. Searching for Life in the Universe: Lessons from Earth. *Annals of the New York Academy of Sciences* 950: 241–258.

Norton, B. 1988. *Why Preserve Natural Variety?* Princeton: Princeton University Press.
Pepperberg, I. M. 1999. *The Alex Studies: Cognitive and Communicative Abilities of Grey Parrots*. Cambridge: Harvard University Press.
Popa, R. 2004. *Between Necessity and Probability: Searching for a Definition of Life*. Berlin: Springer-Verlag.
Putnam, H. 1975. "The Meaning of 'Meaning'." In *Language, Mind and Knowledge*, edited by K. Gunderson, vol. 7 of *Minnesota Studies in the Philosophy of Science*. Minneapolis: Minnesota University Press.
Regan, T. 1983. *The Case for Animal Rights*. Berkeley: University of California Press.
Rothenberg, D. 2008. *Thousand Mile Song*. Philadelphia: Basic Books.
Schulze-Makuch, D., and L. N. Irwin. 2006. "The Prospect of Alien Life in Exotic Forms on Other Worlds." *Naturwissenschaften* 93: 155–172.
Schusterman, R. J., J. A. Thomas, F. G. Wood, and R. Schusterman, eds. 2007. *In Defense of Dolphins: The New Moral Frontier*. Cambridge: Blackwell.
Seeley, T. D. 1996. *The Wisdom of the Hive: The Social Physiology of Honey Bee Colonies*. Cambridge, MA: Harvard University Press.
Seeley, T. D., and P. K. Visscher. 2003. "Choosing a Home: How the Scouts in a Honeybee Swarm Perceive the Completion of Their Group Decision Making." *Behavioral Ecology and Sociobiology* 54: 511–520.
Singer, P. 1975. *Animal Liberation*. New York: HarperCollins.
——, ed. 1993. *A Companion to Ethics*. Oxford: Blackwell.
Slobodchikoff, C. N., J. L. Verdolin, and B. S. Perla. 2009. *Prairie Dogs: Communication and Community in an Animal Society*. Cambridge, MA: Harvard University Press.
Smith, R. M. 1997. *Civic Ideals: Conflicting Visions of Citizenship in U.S. History*. New Haven: Yale University Press.
Tajfel, H. 1969. "Cognitive Aspects of Prejudice." *Journal of Social Issues* 25: 79–97.
——. 1970. "Experiments in Intergroup Discrimination." *Scientific American* 223: 96–102.
Tajfel H., M. G. Billig, R. P. Bundy, and C. Flament. 1971. "Social Categorization and Intergroup Behavior." *European Journal of Social Psychology* 1: 149–178.
Tajfel, H., and J. C. Turner. 1979. "An Integrative Theory of Intergroup Conflict." In *The Social Psychology of Intergroup Relations*, edited by W. G. Austin and S. Worched. Monterey, CA: Brooks-Cole.
United Nations. 1948. *Universal Declaration of Human Rights*. 271 A (III). New York: UN General Assembly.
Ward, P. 2005. *Life as We Do Not Know It*. New York: Viking Penguin.
Warren, M. A. 1997. *Moral Status: Obligations to Persons and Other Living Things*. New York: Oxford University Press.
Weber, M. 2004. "Science as a Vocation" and "Politics as a Vocation." Translated by R. Livingstone. In *The Vocation Lectures*, edited by T. Strong and D. Owen. Indianapolis: Hackett. Originally published in 1917 and 1919.

CHAPTER THREE

Astrobiology and Beyond

From Science to Philosophy and Ethics

William R. Stoeger
VATICAN OBSERVATORY

Astrobiology studies the characteristics and development of cosmic, galactic, stellar, and planetary environments, including those of Earth, insofar as they provide the conditions for the emergence and evolution of life. It also addresses questions about the constitution and possible varieties of life in those environments: What is the range and frequency of life—conscious life, rational, and societal life—in our Universe? These strictly scientific investigations naturally lead us to issues that take us beyond astrobiology and beyond the foundational sciences upon which it relies. What is life? How would the existence and character of extraterrestrial life—and extraterrestrial intelligent life—affect us? What should be our attitude toward it? How should we relate to it, treat it, care for it and for the environments in which it emerges and flourishes? What principles should guide our attitudes toward it and our engagement with it? What ethical demands does it make on us?

Astrobiology presents us not only with scientific, philosophical, and ethical challenges, but also with opportunities in all these spheres. In the practical philosophical and ethical sphere it stimulates us to become more aware of how our relationships, responsibilities, and actual and potential resources extend beyond the local and the present to the global, the cosmic, and the past and future—beyond Earth-bound life and societies to life and societies elsewhere in the cosmos. Thus, astrobiology encourages us to expand and deepen our views of society and self.

Here I shall outline the context and a framework for linking these larger philosophical and ethical reflections with our growing scientific knowledge in astrobiology and in closely related fields. In doing so I shall provide a basis for understanding how important philosophical, social, and ethical concepts like the good, meaning, value, and responsibility are connected with our scientific knowledge—even though they themselves are not scientific concepts. This will help us understand why and how astrobiology, as well as the scientific disciplines on which it depends, can have far-reaching influence on our attitudes, behavior, policies, and projects, both here on Earth and in our cosmic neighborhood. In the best case it can stimulate enhanced social openness, awareness, and responsibility. This influence does not depend on actually confirming or encountering life and intelligence elsewhere. Our growing, strongly supported realization that extraterrestrial life and intelligence are possible, and even likely, broadens and deepens our perspectives, our appreciation, and our sense of connectedness and balanced responsibility for ourselves, for one another, and for our terrestrial and cosmic environments. It can also contribute a new awareness of who we are as knowers and agents in the world and in the Universe.

What is the relevance of astrobiology for ethics, and what is the importance of ethical considerations for astrobiology? Those are the key questions this chapter addresses. Now I shall summarize the conclusions for which I shall argue.

Astrobiology reveals the network of astronomical, chemical, and ecological conditions, relationships, and developments upon which the origin, sustenance, and evolution of life depend, not only elsewhere in the cosmos but also right here on Earth. In doing so it encourages us to understand and appreciate that our highly successful and relatively advanced existence here is connected, and shares basic features, with the physics, chemistry, and almost certainly the biology elsewhere in our Galaxy and in our Universe. Ethics for its part is about recognizing and respecting relationships. It is about how intelligent, freely choosing agents and societies can live and act in harmony with the lower- and higher-level networks, ecosystems, organisms, and communities upon which they depend and in which they exist and function. This sensitivity is deeply consonant with the fundamental ethical insight of one of the most influential early proponents of conservation and ecological policy in the United States, Aldo Leopold (1887–1948). His *Sand County Almanac: And Sketches Here and There* (Leopold 1989) is considered a "charter text" for the movement (Ashley 2010). There Leopold sets forth the basic principle for his "land ethic": "All ethics so far evolved rest upon a single premise: that the individual is a member of a community of interdependent parts . . . The land ethic simply enlarges the boundaries of the community to include soils, waters, plants and animals, or collectively: the land" (Leopold 1989, Ashley 2010). Astrobiology enlarges this perspective beyond our planet to the extraterrestrial systems and communities on which we depend and of which we are a part. In his chapter in this volume, Sullivan uses Leopold's principle in a somewhat different formulation as the foundation of his "planetocentric ethics," and elaborates the ways it can be concretely applied.

Just as biology, geology, paleontology, ecology, and environmental science have increased our understanding of our connections with and our impact on the ecosystems and communities of our planet (Sagan and Margulis 1995), alerting us to our responsibilities beyond just our immediate context and concerns, so astrobiology reinforces that awareness and expands it to embrace the life and environments we may discover and encounter beyond Earth. At the same time it sensitizes us to the challenges and opportunities beyond those that are scientific, technological, and economic. This expands our horizons about who we are and who we may become, and our sense of responsibility to the nearby and the distant communities to which we all belong (see also the chapters in this volume by Cleland and Wilson, and Peters); it also is consonant with the central insights and convictions that are fundamental to Eastern religious and ethical thought, as Irudayadason stresses in his chapter of this volume.

Obviously, then, ethical considerations are fundamental for enabling us to achieve expanded horizons. In this chapter I shall focus on the meta-ethical, or basic philosophical, aspects that underlie all practical ethical policies and decisions (see also the approaches of Bedau, Sullivan, Cleland and Wilson, and Irudayadason in their chapters in this volume). These have to do with what is good and what is bad, what is right and what is wrong, with what has value and what has greater or less value, and with the moral norms or obligations that flow from these conclusions. Determining all of that depends heavily on the overlapping, fundamental communities to which we belong, as Leopold (1989) and Ashley (2010) emphasize, and the important relationships we have with them. This recognition and understanding is essential for developing coherent, universal, and well-supported practical ethical norms, policies, and protocols. This also strongly influences the ways we think about and engage systems, organisms, and communities here and elsewhere. It is clear, for instance, that we must consider possible risks to ourselves and to our terrestrial ecosystems from extraterrestrial organisms. And so we develop precautionary protocols (for detailed discussions, see Bedau, Race, and Sullivan in this volume, and Benner and Woolf's dialogue in the appendix). At the same time, out of concern for the integrity of actual or possible extraterrestrial ecosystems, we find that we need to take definite precautions against infecting extraterrestrial biosystems with terrestrial organisms. There are also ethical concerns related to how we treat, exploit, and commodify extraterrestrial—and terrestrial!—resources, especially if they are alive and possibly conscious and capable of suffering. What moral status do they have—or should we presume they have, before we have adequate knowledge of them? How do we discern the value various environments and life-forms have intrinsically, not just as values for us?

These examples help us see the key connections between astrobiology and ethics—really an extension of the relationship between ecology and environmental sciences and ethics. Again, it is crucial to recognize the networks of relationships and our role and responsibilities as intelligent agents in affecting the functioning and integrity of those networks and the organisms and persons they embrace. With further research and critical reflection on our experience, we can assess the outcomes and even anticipate the consequences of our actual and proposed activities. This will lead us to an understanding of what undermines and what enhances harmony and compatibility with our environment, and with other environments and communities. Such evaluation enables us to fine-tune our ethical norms and protocols.

In the course of this chapter I provide further support and reflection for these conclusions linking astrobiology and ethics, at the level of metaethical foundations.

In the next section I discuss the foundations of ethical thinking and how astrobiology relates to them. From observing and engaging the realities surrounding us, how do we reliably arrive at ethical principles that guide our attitudes and our behavior as individuals and as societies? In the course of doing that, I briefly review some of the key approaches to validating ethical reasoning, and argue for those that are based on knowledge and understanding of the contexts in which, and the relationships through which, we and others thrive. This leads to the emergence of objective ethical criteria that attend to those relationships, to the value of other systems and organisms and the environments they inhabit, and to the actual or anticipated consequences of our active and passive engagement with them (Ladrière 1972)—both terrestrial and extraterrestrial.

In the third section I briefly outline some of the broad cosmological and astrobiological understandings we have of our Universe and several of the general characteristics reality manifests. They provide the overall context within which complex, interacting systems and life emerge. Their character, organization, functional capabilities, and interrelationships with one another and with us reveal the goods and values that underpin the ethical dimensions of reality and guide our engagement with it. In this context I shall briefly review the differentiation and complexification of physical and chemical systems and environments that involve the emergence of networks and levels of evolving relationships. Here the internal and external constitutive relationships among the components of systems and subsystems and with their environments are crucial. I also discuss some of the conditions necessary for the development of complexity and, therefore, life. This provides us with a basis for identifying what is good, and what is of value, which is essential to ethical considerations and ethically motivated action.

Finally, in the fourth section I continue and deepen the discussion, begun in the second section, of how the complex and intricate dynamic order we see around us, and our personal and communal participation in and reliance on it, leads us to develop the important concepts of goods, meanings, values, obligations, and responsibilities. These amount to ethical and more broadly to philosophical and even theological considerations. Several simple and obvious examples are mentioned, with the view to demonstrating how astrobiology is expanding, modifying, and deepening our views of ourselves, society and the world, and the Universe—and our appreciation of how we are to fruitfully and responsibly engage them.

Ethical Thinking and Astrobiology

The fundamental ethical principle is, "Do good, and avoid evil." But how do we determine what is good and what is evil? And once we do that, how do we establish priorities and choose among the different goods, values, responsibilities, and obligations we experience as individuals and as communities (Fagothey 1989)? This is difficult enough with regard to our relationships with people, communities, and environments of which we are aware and with which we are immediately concerned. How can we go about fulfilling this with communities, environments, and life-forms that are neither "like us" nor directly connected to us—those here on Earth and those we may discover in another part of the Solar System or the Galaxy?

As an example, we shall consider "life." Most people agree at some level that life is good, is a value, and should be preserved. It has some definite level of ethical or moral status (Rolston 1995). Certainly McKay in his contribution to this volume argues strongly for this position, as do Cleland

and Wilson in their chapter and Sullivan in his. But what is the basis for giving life value and moral status? Furthermore, it is clear that not all life has the same status. We consider human life privileged—particularly human life like ours. But human life that is not like ours or is a threat to us or nonhuman life is usually considered to be of much less value—to have a much lower ethical status. Thus, our ethical priorities value other human beings more than lower, or different, forms of life. And the care and respect we give to human life and society trump that we give to other biotic environments and to environments devoid of life. What is the justification for such differences in priority and privilege? Ostensibly it is because we see life as a genuine good, and we value it, and further we value higher forms of life, particularly intelligent life, over less-developed kinds of life (Fagothey 1989; Nozick 1989). We need to delve more deeply and critically into the basis for these preferences and the values and ethical norms that flow from them, and then relate them to the attitudes and relationships that will govern our engagement with extraterrestrial life and intelligence.

Before doing that, however, we need to look briefly and broadly at approaches to moral reasoning. In doing so, I first leave out those that are primarily religious. I do that not because they are unimportant, but simply because there are strong philosophical foundations for ethical principles and norms that do not rely on religious belief. Some of these purely philosophical approaches have been elaborated and bolstered by theological or faith-based considerations. Later I briefly comment on several additional supports to ethics religion and theology often provide. But at this stage it is important to realize that much of ethical reasoning and development relies on philosophical, not theological, considerations.

There are a variety of competing schools in the foundations of ethics—or what some refer to as "meta-ethics." There are subjective, utilitarian, emotivist, virtue-based, and various objective approaches to ethics (see the latter part of Cleland and Wilson's chapter in this volume for their treatment of utilitarianism and duty ethics). I cannot survey these here. But for our purposes I briefly discuss two of the primary fault lines among them. Many deny that there are any objective and verifiable bases for ethics, because ethical principles and norms cannot be reduced to testable, purely scientifically supported statements. However, many other influential and compelling ethical approaches, while recognizing that ethics is not science (see Cleland and Wilson, this volume), strongly affirm that ethics does rely in large part on the way reality is organized and, therefore, must be informed by scientific understanding of the world and the cosmos (Fagothey 1989; Pojman 1999; Ayala 1995). Here I espouse this second course—because it is objective, is very insistent on obtaining all the relevant information, including that from the sciences, and leads in the long term to a form of verification—via the resulting good or harmful consequences of a course of action or policy implementation.

The second closely connected long-standing controversy, which exemplifies the substantial differences between approaches to ethical foundations, has to do with what is known as the "naturalistic fallacy" (Pojman 1999; Fagothey 1989; Cleland and Wilson, this volume). This states that values and obligations follow from "facts," often expressed in this way: "From what is, we can determine what ought to be done." It has often been referred to as a "fallacy" because many philosophers and ethicists, beginning with Hume, have rejected it, maintaining that one cannot go from facts to values, "from 'is' to 'ought.'" They strongly insist on this prescription, even to the extent of denying any relationship between what is and what we ought or ought not to do. However, many others—especially more recently—have embraced some form of naturalism. Though most admit that we cannot reduce ethical norms and obligations directly to scientifically verifiable facts, they strongly affirm that ethics does depend heavily on the way things are (Fagothey 1989; Ayala 1995; Pojman 1999). Here we take this modified naturalistic position, principally because it takes seriously all that we know from the natural sciences, as well as our ability to act and engage reality constructively. The key point here is that the irreducibility of values or "oughts" to what is derives from the recognition and assignment of goodness and value to an aspect or component of what is by a rational self-reflective agent. This assignment of goodness or value depends heavily on what is, and can be rationally and critically evaluated and questioned in light of the facts, but is not reducible to them. This is why my position is a "modified naturalism."

This leads us to the key issue we raised before discussing meta-ethical disagreements: How, from our naturalistic perspective, can we decide or determine what is good and what is less good or bad? It is on the basis of that recognition that we begin to see what we are morally allowed or required to do or not do, and what norms and moral policies we should adopt as individuals and as a society. Other related issues rest on our identification of the good: what is of value, what is of less value, what we ought to do or not do. Of course, there are many things that are good that we do not do and are not required to do. They are optional. Many of our actions as individuals and societies are morally indifferent. But some actions are required, and some are definitely to be avoided. I return to a fuller discussion of the connection between what is good and what is of value and what is of obligation in the fourth section of this chapter. I concentrate here on examining the criteria for calling something good.

Essentially, what is good depends on its relationships with other things and how it fits or would fit within a system, an organism, a person, a community or society. Is it compatible with the relationships that maintain the system, organism, or community? Does it support it, enhance it, or instead does it threaten, compromise, or diminish it? What are the actual or anticipated consequences of performing an action on a system or a community? What is beneficial and supportive is good and, therefore, is of value. What is harmful to it is not. Acting in harmony with our environment, and with the networks, systems, and relationships that support us and provide us meaning and direction, is an overarching good.

As Pojman (1999), a well-known proponent of the modified naturalistic point of view, says, ethics or morality has a definite point to it, "it has to do with producing better outcomes—in terms of happiness, flourishing, welfare, equality and the like. . . . A certain logic pertains in what can be called morally good, depending on non-moral values" (233). Thus moral or ethical properties, though not independent properties or "facts," are functional, "standing for practices which tend to fulfill the purpose of morality (the right) or tend to thwart it (the wrong)" (244). All practices, behaviors, and policies are directed toward achieving certain ends (Ladrière 1972). Of course, the further questions are: Which ends or purposes are we talking about? What values are we privileging? And what is the price that we and others (including the larger community, global society, and the environment) must pay for pursuing those ends? Obviously, if the ends or the outcomes are destructive and disruptive of the communities and environments of which the conscious freely choosing agent is a part, then that activity will be unethical—even if it satisfies the agent's immediate desires or ends.

As we indicated earlier, this perspective obviously confirms and expands the ethical foundations Leopold (1989) has identified and emphasized in his "land ethic"—that we all belong to overlapping communities that extend outward, into the past, and into the future, upon which we are completely dependent and which, to varying degrees, are dependent on us.

These criteria are complemented by the overarching criterion for ethical acceptability suggested by Kant—his categorical imperative (Rogerson 1991; Pojman 1999). An action, policy, or attitude is ethically right only if it can be universalized. If you cannot consistently will that everyone act in the way that you are considering—because it would yield harmful, disruptive, destructive, or contradictory outcomes—then that action or policy is unethical. This indirectly reveals what is in harmony with the nature and proper functioning and survivability of systems, environments, and communities, and what is not. A related criterion is the golden rule, "Do unto others what you would have them do to you" (Rogerson 1991). At least from our experience, it seems clear that this applies only to consciously aware, rational, and freely choosing beings.

As with all general ethical principles, the application of the golden rule and the categorical imperative must be modulated by considering the concrete circumstances, and conflicting choices of rights, goods, and values, that usually attend any actual situation—for instance, in cases of self-defense against unjustified attack.

There is a subtlety here that needs to be recognized. Sometimes what appears to be pleasurable, beneficial, or supportive in the short term may be ultimately harmful in the long run. In other circumstances what seems to be painful, unpleasant, and harmful in the short term may be very beneficial and helpful in the long run. Our recognition and assignment of goodness and value are relative to our knowledge and understanding and to our intentions and purposes. There are innumerable examples of this with which we are all familiar. Liberal use of fossil fuels and other chemicals supports and enhances life and production now, but may lead to severe problems later—the effects of pollution and global warming, for instance. A definite good for a particular subsystem, person, or small group may undermine and compromise the larger systems and communities of which it is a part. Of course, the opposite situation may also occur. What is good for the larger system may be very harmful for certain components or subsystems. Thus, whether something should be considered good or not depends on looking carefully at the long and the short term, the purposes the agents have in mind, as well as the local and the more global consequences, and the levels of values and moral status we attach to various entities and the systems they inhabit.

Earlier I promised to mention the impact of religion on ethical reasoning. Clearly there are always moral obligations deriving from religious belief and commitment. These as all religious beliefs are subject to discernment, critique, and ongoing refinement in light of experience to determine their authenticity and legitimacy. Do they contribute to the healthy development, inspiration, and enhancement of the community, and the larger society and environment of which it is a part? Genuine religious beliefs and insights will usually confirm the conclusions of correct ethical reasoning. The motivations and justifications for doing so, however, may often go beyond those philosophy and the sciences provide. For instance, belief that God has created all things is often cited as a reason for respecting, caring for, and sharing our resources. And the religious conviction that human beings are made "in the image and likeness of God" supports the moral status and priority we give to them.

Now we briefly look at the evolving order and complexity that characterize the Universe and so many of the interrelated systems that have emerged within it. Understanding and appreciating them is fundamental to valuing and living in harmony with them.

The Evolution of Our Life-Bearing Universe

Our Universe, which now manifests so many different levels of organization and complexity and within which life, consciousness, societies, and cultures have emerged, began as a very simple, expanding, cooling system of gas and radiation (for a brief, clear summary of cosmic history, see, for instance, Kolb and Turner 1990). It was almost perfectly smooth in density for the first several hundred thousand years after the Big Bang. Until stars formed and began synthesizing heavier chemical elements, it consisted only of hydrogen, helium, and a trace of lithium, the lightest metal. In its beginning about 13.7 billion years ago, our Universe transitioned from a highly energetic quantum cosmological state, called the Planck era, with a temperature $T \geq 10^{32}$ Kelvin, and expanded very rapidly under the influence of dominant vacuum energy in what is known as the "inflationary era." During that very brief period it expanded by a factor of at least 10^{29}, and was endowed through quantum fluctuations with the primordial density seeds for galaxies and clusters of galaxies. Very soon afterward, the vacuum energy decayed and was transformed into particles and radiation, thus reheating the Universe. From that time onward—all well within one second after the Big Bang—it expanded and cooled much more slowly as a nearly homogeneous system for the next 300,000 years or so.

At that point the matter of the Universe decoupled from the radiation. The Universe continued to expand and cool, but now structure formation began in earnest, under the pervasive influence of local gravitational fields, resulting in the formation of galaxies, clusters of galaxies, and stars within the galaxies. With the advent of stars, elements heavier than helium and lithium were produced, 92 elements in all, plus evanescent transuranic elements. These elements eventually spread throughout the cosmic medium as the result of supernova explosions and gas outflows from evolved stellar systems.

In cooler cosmic environments, such as the cool atmospheres of stars and gas clouds, and on asteroids, planets, and comets, uninstructed chemical evolution proceeded, as simple and complex molecules formed in the chemically enriched gas and dust through simple chemical reactions (Küppers 1983). Much later, in some special places (at least on Earth), instructed chemical evolution began, as systems of information carrying molecules (RNA, DNA, proteins) manufactured an increasing variety of large, highly specialized molecules in precisely information-controlled processes (Küppers 1983, 1990). Natural selection began to function at this prebiotic stage. Soon afterward networks of these molecules developed into living cells, and biological evolution began—with all that has flowed from it (see, for instance, Eigen 1992; Smith and Szathmáry 1995; Lunine 2005), via natural selection with the help of genetic mutations, lateral gene transfer, symbiogenesis, sexual recombination, and so forth, during the last four billion years on our planet—and possibly elsewhere.

Finally, with the emergence of consciousness and rationality, we have entered the stage of cultural evolution with the dominance on Earth of humans and their social, political, economic, technological, and religious structures, and the ideas, orientations, interests, and behaviors that flow from them. Thus, cosmic evolution establishes the conditions and the building blocks for chemical evolution. Uninstructed chemical evolution sets the stage for instructed chemical evolution, which in turn enables the emergence of life and the processes crucial for biological evolution. What the evolutionary development of biological organisms has eventually produced here on Earth is a class of beings capable of self-reflective consciousness, intelligence, technology, and culture. Clearly that would have been impossible without the accomplishments of all the earlier stages of evolution and development. There is a deep and highly differentiated connectedness between all that we see in our Universe.

A startling realization of astrobiology is the fact that, although we have very reliable evidence that the laws of physics and chemistry that hold on our planet also hold throughout the rest of the observable Universe, we as yet have no direct knowledge of biology anywhere else. We can, of course, surmise that whatever life has emerged elsewhere, it must be based on the same principles of physics and chemistry. But that leaves open an enormous number of possibilities, whose detailed structure and long-term development we are incapable of adequately understanding or even imagining.

What are the key general characteristics of nature and the Universe that have been revealed by scientific inquiry (Stoeger 2007b)? Certainly, change—evolution on all scales—must be considered one of those. We can also add order and intelligibility and the relative transience and fragility of everything within the Universe.

There are three features that are central to our discussion. The first and most fundamental is *relationality*, highly differentiated and many-leveled interlocking relationality. These relationships among components at each level often constitute the objects and systems at the next-higher level, such as atoms in particular relationships forming molecules with very different properties from those of the atoms that make them up. At other times these relationships link systems with the larger environment in very important and essential ways, protect them from undue intrusions from the outside, or help them adjust and maintain themselves in the face of fluctuating environmental conditions. In biological systems there are many levels of such relationships, which build on and modulate one another.

These levels of interacting, highly differentiated, and evolving relationships enable the second important feature of reality: nested levels of *complexity*—the hierarchical structures of physical, chemical, biological, and sociological/cultural reality. Each lower level is essential, a necessary condition for the higher levels, but interestingly enough, rarely, if ever, a sufficient condition. In the interactions within and among levels, there are obviously crucial bottom-up causal influences, in virtue of that necessary, causal structure. But there are also clearly same-level and top-down causal influences (Scott 2007; Auletta et al. 2008) that constrain, channel, select, coordinate, and enhance or mitigate the lower-level and same-level causal factors.

Then, thirdly, there is what some have referred to as the formational and function integrity (the relative autonomy) of nature (Van Till 1986). Given its existence and its intrinsic order and dynamism, it acts on its own, without need for outside stimulus or initiation. It manifests a deep but highly differentiated *unity*, which is due to the various levels of relationships highlighted above. Nothing falls outside this unity; everything we know is a part of it and related to it in some way.

Two other features of reality we can include in our list are *directionality* and apparent *fine-tuning*. Though we cannot say anything about teleology (design or purpose in nature) from the point of view of the natural sciences, it is clear that there are nested directionalities in nature (Stoeger 1998). Not a unique direction toward one specific goal or end point, but rather a definite range of allowed outcomes from one system configuration to another. Cosmologically there is an overarching directionality toward the cooler, lumpier, and more complex, which is dictated by the expanding, cooling Universe. With any system, considered at a particular point in its history, there is only a certain set of outcomes toward which it can evolve, given the regularities it obeys and the conditions that affect it. A cooling gas cloud that has begun to collapse will behave in only a certain limited number of ways as it forms a cluster of interacting stars.

The apparent fine-tuning of the Universe and of nature is also a surprising and potentially significant feature (see Stoeger 2007a and references therein). If the key parameters describing the laws of nature and the initial conditions of our Universe had been even slightly different, there would be no complexity and no life. If, for instance, the Universe had been expanding either too quickly or not quickly enough after the inflationary era, stars would not have formed, leaving the Universe permanently and severely chemically impoverished, lifeless, and very boring.

Differentiation is a key step toward complexity in the early Universe or in any system. When a system breaks up into subsystems, whether it be a galaxy, a star and planets, or specialized subsystems in biology, some subsystems will be or will gradually become different from others and will therefore evolve differently. Furthermore, the ways in which those entities and subsystems interact, combine, and relate to one another become very different, thus providing enriched environments for further combination and diversification. That is particularly clear with regard to the particles and the stable nuclear combinations they have formed, which in turn have provided nature with the basis for the formation of innumerable functional molecules with an incredible diversity of properties. This chemical enrichment has also strongly affected stellar and galactic evolution, combining with gravity and angular momentum to enable the emergence and evolution of cool, chemically rich venues in which prebiotic uninstructed and instructed chemical evolutionary processes have flourished.

The striking reality is that diversity emerges from these underlying unities. Substantively different relationships develop within these systems. Thus, their dynamic unity encourages diversity eschewing uniformity, and diversity modulates and enhances unity and the dynamic order upon which it relies.

That is only the beginning! In the history of complex systems and life on Earth, there have been innumerable levels of such differentiation and complexification, often followed by selection of the systems or organisms better adapted to a given context or environment. Along with these general processes, there is obviously a specialization of operations and functions of the constituents and the subsystems they make up at each level. That is, each constituent or component subsystem in its interaction or relationship with the larger whole and with its environment makes a specific contribution to and provides a precise outcome for the larger overall system of which it is a part. In many cases the larger system could not function or survive without that contribution or subsystem outcome. Each subsystem fulfills an essential role for the larger system or systems (Stoeger 1998, 2011). In some sense we can even say that the emerging functional "goal" or "reason" for which that component is present in the larger system is to provide that essential contribution (Auletta et al. 2008). In the animal body, for instance, we have the heart, lungs, kidneys, and other organs to provide functions essential for the life of the whole body.

Thus, every system and organism and the Universe itself becomes a differentiated unity. Each system has different subsystems functioning within it, which together give it the properties and capabilities it manifests. And each system finds itself within a larger system upon which it depends and with which it interacts in some way. The reality of which we are a part is really composed of an intricate complex of systems within systems or networks within networks, where some of those systems and networks are communities of persons, persons themselves, and different types of living organisms. In each case there is a whole range of different relationships and interactions that link those systems, organisms, and subsystems together in ways crucial for the proper functioning and continued survival and development of each one, and of the whole complex.

I have already stressed the importance of these relationships and particularly the network of internal and external relationships relative to each entity, system, or organism that is constitutive of its identity—of its being what it is (Stoeger 1999). At every level of organization there are identifiable objects and organisms that have a definite persisting unity and integrity, distinctive properties, capabilities, and behavior. Each is what it is because of the networks of essential and defining relationships among its components, and with its antecedents and its environment, which makes it what it is. This is true of each person, each organism, each molecule, each atom, each star, each galaxy. We live in a profoundly interrelational Universe. Without those innumerable layers of differentiated relationships, nothing in our Universe would be possible or sustainable. We might say that because of those pervasive relationships, reality is constituted by communities nested within communities. This scientifically based understanding and recognition of all these essential, intricately networked interrelationships and interdependencies provides incontrovertible support for Leopold's (1989) and Ashley's (2010) ethical principles, to which Sullivan strongly subscribes in his "planetocentric ethics," presented in his chapter in this volume. It also gives robust support to the profound Buddhist and

Hindu perspectives on the fundamental unity and interconnectedness of all things in the Universe, which Irudayadason discusses in his chapter.

At the level of instructed chemical evolution, biology, and consciousness, these defining relationships become even more precise and particular. At those advanced levels of organization, detailed information is generated, selected, preserved, faithfully reproduced, and implemented, as systems and organisms interact with their environment and natural selection sifts the more suitable from the less suitable (Küppers 1983, 1990). In these cases there is a sophisticated genetic "memory" through which important information from the past is accumulated over many generations and passed on to descendants (Deacon 2006, 2007). Furthermore, mechanisms of variation function to explore and stabilize new and more successful genetic combinations through which evolution occurs and new information is added to the memory pool.

Before moving back to philosophy and ethics, it is worth briefly cataloging some of the principal conditions that must be fulfilled if complexity and, a fortiori, life are to develop. One of the preliminary and often unmentioned prerequisites is "locality." For complexity to even have a chance of developing, there must be the relative isolation of local regions in spacetime from global or long-range effects and disturbances. Once this has been accomplished, separation, actual condensation, and encapsulation of structures, systems, and subsystems are essential. This allows for each system's persistence and evolution as a separate entity, but also for its controlled and mutually beneficial openness and interaction with other systems and with the surrounding environment upon which it depends. In all these cases there will be bottom-up, same-level, and top-down causal influences (Scott 2007; Auletta et al. 2008). These will differ in character but will nevertheless be both mutually compatible and operative.

Other requirements for the development of complexity, implied or presaged by those we have already mentioned, are these:

- Cooperative and symbiotic behavior and the synthesis of more basic systems into higher-level systems with consequent emergent properties and behavior, such as atoms into molecules.
- Feedback control systems, effective transmission and implementation of information, and goal-like or teleonomic behavior (Ayala 1970; Stoeger 1998; Auletta et al. 2008).
- Modularity and redundancy (Ellis 2006)—development of subsystems whose inner workings are screened from the larger system but whose products and contributions to it are essential or duplicate or

substitute for the functionality of other systems, providing an important backup and thus enhancing the long-term viability and flexibility of the overall system or organism.

- In living systems, the generation, selection, preservation, accumulation, reproduction, and effective and coordinated implementation of information.

These conditions make natural selection, and therefore biological evolution, feasible.

With this review of the relational and intricately interaction-ordered character of nature in mind, we return to continue our discussion of metaethics, which provides a secure basis for moral decision making. These ethical foundations, which are applicable to ecology and astrobiology, depend upon and derive in large part from the deeply interrelational character and nature of physical, chemical, biological, and personal reality as the sciences and other channels of human experience reveal them.

From Order and Differentiated Unity to Goods, Meaning, and Values

Earlier I emphasized the highly differentiated unities of systems and subsystems, which make up the Universe and all of reality. Each of these manifests many different interacting levels of networks and relationships and a dynamic and effective order, which have been provisionally modeled by the natural sciences.

One of the key transitions beyond the sciences to philosophy and ethics is our perception and evaluation of this order, and these networks and levels of relationships, as goods. This means that we judge that order, those relationships, and those systems, organisms, and environments as positive, relative to their absence, and relative to one another and to ourselves, and therefore as beneficial and in need of protection and preservation. They are goods in themselves, because of their existence and intrinsic organization, complexity, constitutive relationships, and beauty. But they are also goods because of their relationships with other things—because of the essential and nonessential contributions they make to the systems, organisms, and ecologies of which they are a part.

These goods, then, have value—both intrinsic value, which derives from the degree of each one's organic unity (Nozick 1989), from its constitutive relationships and its intrinsic actual or potential capabilities, and extrinsic value, which is due to its importance, usefulness, and contributions to another system, organism, person, or community. (See also the chapters by Bedau, Cleland and Wilson, Race, and especially Sullivan in this volume. As Irudayadason, this volume, stresses, Eastern ethical thought also strongly emphasizes the intrinsic value of all things.) Either way, these values naturally demand that we take a definite stand (action) with regard to the goods we encounter, that we engage those realities positively in a variety of ways (see especially Sullivan, this volume, for concrete policies and their practical implementation), depending upon their intrinsic characteristics and their relationships with other systems, but also upon our interests, our capabilities, and our personal and communal priorities. These goods have become values, which Bernard Lonergan (1970) pragmatically defines as goods considered as objects of possible rational choice—which can be on account of either the intrinsic properties and characteristics of the object, or its importance and utility to the potential chooser, or both. Values in this sense undergird, motivate, and guide all human individual and communal decision making and behavior insofar as they are rational.

Thus, a number of key values are assumed and operative in scientific and cosmological research itself, such as the values of truth, understanding, and knowledge; of order, elegance, and simplicity; of coherence and unity in the midst of diversity; of the success of hypotheses. These are not often specifically adverted to as essential values in doing science, much less probed for their origin and justification. That is the role of philosophy and philosophy of science. Those basic characteristics and values take us beyond the interests and competencies of the sciences themselves.

Finally, from certain personal and social appropriations and applications of values, "oughts" obligations, or "objective requirements" often arise (Rokeach 1973). These are demands that our personal and social interpretations of reality impose upon us if we are to be in harmony with the communities, systems, and environments to which we belong. Then there is meaning! Meaning is just as important as value, serves to complement it, and supports our value-based decisions. Perception of value without meaning is personally and communally empty. Meaning personalizes value, and deepens it. Meaning depends on the relationships we have with something, or someone else—and the relationships the things and people around us have with the larger, encompassing wholes or systems to which they belong (Nozick 1989) or more directly perhaps to a more fundamental reality—to a particular outcome, or to the relationships that connect us to those larger or more fundamental realities. However, as individuals and communities we often are blind to or unaware of many of those larger, more encompassing systems, communities, and environments, even though we depend upon them. Thus, one of the paths to expanding and deepening, or possibly recovering, our values and meanings is by informing and sensitizing ourselves to the larger, richer world and communities around us, even those beyond our planet and Solar System, and to the connections, dependencies, and commonalities we have with them.

Both Leopold (1989) and Ashley (2010) confirm the importance of this kind of personal, ethical engagement. Ashley comments, "For Leopold, it is by attending to our surroundings that the circle of solidarity can widen, that a broader compass of beings can come into view that make a claim on us in various ways" (Ashley 2010, 27). Ashley continues: "[Remembering and recognizing the Earth's species that are today threatened] challenges us to let those species have a claim on us, to be included in some way in the circle of solidarity that defines those who will count in our moral evaluations. This . . . does not depend on absolute symmetry: we need not attribute humanity or 'subjecthood' to the other members of the biotic community for them to have a claim on us. As Leopold suggests, our capacity to let the other make a claim on us when that other is unlike us . . . is part of what makes us distinctly human" (29–30). This succinctly and forcefully expresses the essential foundation for a robust ethics, including its extension into the realms of ecology and astrobiology. It also resonates very clearly with the foundations of Buddhist and Hindu ethical thinking (see Irudayadason, this volume). Our recognition and parsing of the goods, values, and oughts based on the networks of relationships we enjoy with the various systems and communities to which we belong serve to elaborate, strengthen, and explain those foundations.

Figure 3.1 summarizes the structure of our interactions with reality. On the left side is reality as it is in itself (including all knowers and doers). That is the ontological side of things. On the right side is epistemology, the results or products of the knowers' and doers' active rational and personal engagement with that reality. The interaction of knowers and doers with reality generates "perceived reality"—which is reality as we perceive it, experience it, know it, and model it. Perceived reality includes the understandings, perspectives, and knowledge gained by the various sciences, arts, humanities, schools of philosophy, and religion. More basically it critically embraces the "facts," meanings, values, oughts, and relationships that result as knowers and doers interact with reality on a daily basis in many different prescientific and predisciplinary ways. It is obvious from this diagram and from our broader reflections on our experiences that, although

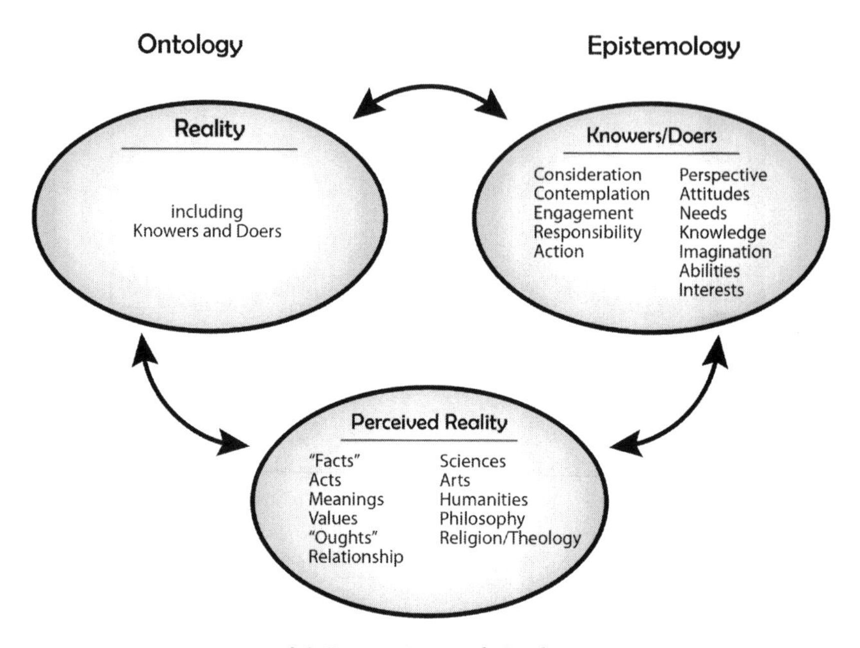

3.1 Interactions with Reality

physics, cosmology, biology, and astrobiology can and do make important contributions to the construction and continual purification of our perceived reality, they are and need to be complemented by a number of other important sources of reflection, knowledge, and understanding.

Figure 3.2, a modified version of a diagram by Calvin DeWitt (1998), indicates the mutual influences among principal components of human culture that determine our choices, our behavior, and the results stemming from them. DeWitt's original diagram is the inner triangle linking science, ethics, and praxis (defined as action or practice). Ethics needs to take scientific data and conclusions into consideration, but science in turn must consider ethical constraints, and both ethics and science contribute to guiding praxis. Our reflections on the success and influence of our praxis will often force some modification in our ethics and in our science (observation, experimental conclusions, and applications). I have added "Experience" below the inner interaction triangle, because everything above that "Experience" line is the result of organized critical reflection on experience—and all that is above the line enables our awareness and articulation of experience. I have added "Philosophy and Theology" above the inner triangle, because these disciplines either directly or indirectly

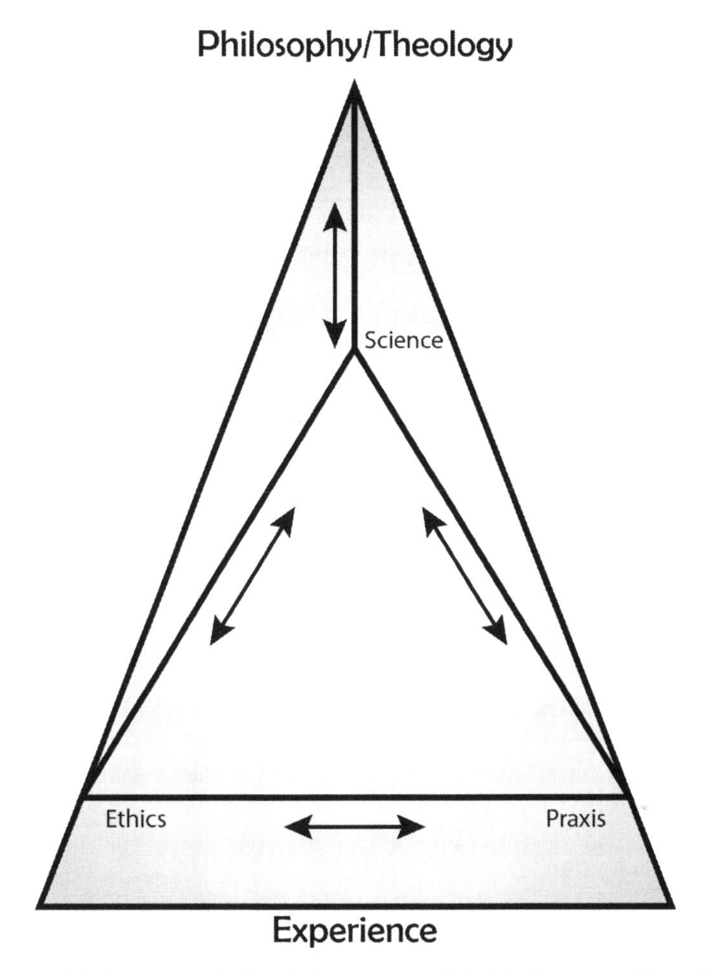

3.2 Mutual Influences of Cultural Components. Modified from DeWitt 1998.

influence ethics, science, and praxis. Their very different influences on ethics and on science are clear. Certainly, too, an individual's religious and philosophical positions will strongly affect his or her way of acting and living. That is true of communal acting and living as well.

One of the main challenges we have as individuals and as societies in this regard is prioritizing values and oughts. As humans we manifest a number of core values in individual and social life, but we often have trouble ordering them and agreeing on which ones should take priority in given circumstances. Thus, the key values we agree on are often experienced to be in conflict with one another. Ethics deals with these values and the "oughts" that flow from them, and with how conflicts among them in concrete circumstances can be resolved with the prioritization of values and their proper expression.

My main focus here has been to provide in a very simple and straightforward way the larger cosmological and natural context within which consideration of goods, values, oughts, and meanings naturally emerge, and how these take us beyond what the sciences directly treat to philosophical and ethical discourse, which in turn leads to action—to praxis. The sciences themselves, as we have already mentioned, presuppose some of the key values, but they do not explicitly reflect on them as objects of their methods and their interests.

The relevance of these reflections to astrobiology is obvious. As inquisitive and scientifically oriented human beings, we consider knowledge and understanding of the Universe as a good with intrinsic value. We have come to consider other planets, moons, asteroids, comets, organisms, and ecologies also as goods with intrinsic and possibly some extrinsic value, and having some meaning, particularly in terms of their indirect and potentially direct connections with ourselves. We share the same cosmic origins, and for the life we may find in our own immediate environs, the Solar System, the same stellar and planetary system origins. All we see is connected however indirectly. And in the future we may interact in a much more direct way. We are capable of influencing and affecting extraterrestrial life and both biotic and nonbiotic environments, and they are already affecting us and have the potential to affect us even more deeply.

Furthermore, as we have already seen, because of the order and ecology it would manifest, its local biosphere, and the potential it has for evolving, any extraterrestrial life possesses intrinsic value and therefore demands some definite ethical consideration from us. How those considerations should enter our priorities is another question, of course, which requires further evaluation and discussion. At the very least they compel us to realize and respect our deep and often neglected relationships with the larger communities of which we are part. They expand our awareness of and interest in other places in the Universe where order, complexity, and probably life and consciousness are thriving. At the same time we begin to appreciate also what a special place Earth is, its rich but very fragile and sensitive ecology, and both its vulnerability and its future potential. Thus we can see that astrobiology expands our views of society and self in a very direct way.

Before closing I turn to several concrete examples of how philosophical and ethical issues have functioned and are functioning in ways parallel to those arising in astrobiology. Of course, the first and most obvious one we have just mentioned is that of our increased understanding of our ecology and our environment and the limits to our resources: How must we modify our lives and our work so as to live on this planet in a sustainable and responsible way?

The second example is the European encounter with the cultures of the New World in the fifteenth and sixteenth centuries. Among the questions that arose were these: Were these newly discovered societies human? How were they to be treated? Peters discusses these issues in depth from an astrobiological point of view in his contribution to this volume.

The third example is really a generalization of the second: How do we, or should we, engage "the other," whether living or not, overtly rational or not? As a threat, as a potential commodity, as a stranger, as a potential companion? And on what grounds do we decide and act upon our decisions? (Again, see Peters's chapter in this volume for an extensive consideration of these questions.)

We are self-aware and intentional agents of novelty and change, both as individuals and as communities. We are aware of the goals for which we act, and are capable of anticipating the consequences of what we do. At the same time we are growing in our recognition and understanding of the interlocking levels and networks of the physical, biological, psychological, and sociological order and complexity in which we participate. Some of these are personal and local. Others that overlap and sustain them are much more extended and radically social. And still others are international, global—and even extraterrestrial and cosmic. As our scientific, technological, and social reach expands more and more into these frontiers, our ethical sensitivities and concerns must follow. They do so by honoring and valuing each system, network, organism, and environment, giving priority to those that are more complex and responsive—to life and especially to conscious, rational life.

This implies, of course, that we also value and protect the integrity and development of the environmental and social ecosystems that sustain reproduction, growth, and evolution. In the case of uncertainty, we must err on the side of caution and care.

Thus, we find ourselves responsible for fostering and ensuring the health and harmony of the overlapping environments, contexts, and communities within which we live, and with which we interact or will interact—whether Earth-bound or cosmic. In particular, as we anticipate and invite contact with extraterrestrial life, we continue to carefully consider the various ways it may affect us, positively and negatively, as well as how we in turn can and should responsibly and ethically interact with it.

References

Ashley, J. M. 2010. "Reading the Universe Story Theologically: The Contribution of a Biblical Narrative Imagination." *Theological Studies* 71: 870–902.

Auletta, G., G. F. R. Ellis, and L. Jaeger. 2008. "Top-Down Causation by Information-Control: From a Philosophical Problem to a Scientific Research Program." *Journal of the Royal Society Interface* 5: 1159–1172.

Ayala, F. J. 1970. "Teleological Explanation in Evolutionary Biology." *Philosophy of Science* 37: 1–15.

———. 1995. "The Difference of Being Human." In *Biology, Ethics and the Origins of Life*, edited by H. Rolston III. Boston: Jones and Bartlett.

Deacon, T. W. 2006. "Emergence: The Whole at the Wheel's Hub." In *The Re-Emergence of Emergence: The Emergentist Hypothesis from Science to Religion*, edited by P. Clayton and P. Davies. Oxford: Oxford University Press.

———. 2007. "Three Levels of Emergent Phenomena." In *Evolution and Emergence: Systems, Organisms*, edited by N. Murphy and W. R. Stoeger. Oxford: Oxford University Press.

DeWitt, C. 1998. "Science, Ethics and Praxis: Getting It All Together." In *Ecology and Religion: Scientists Speak*, edited by J. E. Carroll and K. Warner. Quincy, IL: Franciscan Press.

Eigen, M. 1992. *Steps towards Life: A Perspective on Evolution*. Oxford: Oxford University Press.

Ellis, G. F. R. 2006. "On the Nature of Emergent Reality." In *The Re-emergence of Emergence: The Emergentist Hypothesis from Science to Religion*, edited by P. Clayton and P. Davies. Oxford: Oxford University Press.

Fagothey, A. 1989. *Right and Reason: Ethics in Theory and Practice*, an updated version revised by Milton A. Gonsalves. Columbus, OH: Merrill.

Kolb, E. W., and M. S. Turner. 1990. *The Early Universe*. New York: Addison-Wesley.

Küppers, B. O. 1983. *The Molecular Theory of Evolution*. New York: Springer-Verlag.

———. 1990. *Information and the Origin of Life*. Cambridge, MA: MIT Press.

Ladrière, J. 1972. *Language and Belief.* Translated by Garrett Barden. Chap. 5. Dublin: Gill and Macmillan.

Leopold, Aldo. 1989. *A Sand County Almanac: And Sketches Here and There*. New York: Oxford University Press.

Lonergan, B. J. F. 1970. *Insight: A Study of Human Understanding*. New York: Philosophical Library.

Lunine, J. I. 2005. *Astrobiology: A Multidisciplinary Approach*. San Francisco: Pearson/Addison-Wesley.

Nozick, R. 1989. *The Examined Life*. New York: Simon and Schuster.

Pojman, L. P. 1999. *Ethics: Discovering Right and Wrong*. 3rd ed. Belmont, CA: Wadsworth.

Rogerson, F. J., ed. 1991. *Introduction to Ethical Theory*. Fort Worth, TX: Holt, Rinehart and Winston.

Rokeach, M. 1973. *The Nature of Human Values*. New York: Free Press.

Rolston, H., III, ed. 1995. *Biology, Ethics and the Origins of Life*. Boston: Jones and Bartlett.

Sagan, D., and L. Margulis. 1995. "Facing Nature." In *Biology, Ethics and the Origins of Life*, edited by H. Rolston III. Boston: Jones and Bartlett.

Scott, A. 2007. "Nonlinear Science and the Cognitive Hierarchy." In *Evolution and Emergence: Systems, Organisms, Persons*, edited by N. Murphy, and W. R. Stoeger. Oxford: Oxford University Press.

Smith, J. M., and E. Szathmáry. 1995. *Major Transitions in Evolution*. San Francisco: W. H. Freeman.

Stoeger, W. R. 1998. "The Immanent Directionality of the Evolutionary Process and Its Relationship to Teleology." In *Evolutionary and Molecular Biology: Scientific Perspectives on Divine Action*, edited by R. J. Russell, W. R. Stoeger, and F. J. Ayala. Vatican City State and Berkeley, CA: Vatican Observatory Publications and the Center for Theology and the Natural Sciences.

———. 1999. "The Mind-Brain Problem, the Laws of Nature, and Constitutive Relationships." In *Neuroscience and the Person: Scientific Perspectives on Divine Action*, edited by R. J. Russell, N. Murphy, T. C. Meyering, and M. A. Arbib. Vatican City State and Berkeley, CA: Vatican Observatory Publications and the Center for Theology and the Natural Sciences.

———. 2007a. "Are Anthropic Arguments, Involving Multiverse and Beyond, Legitimate?" In *Universe or Multiverse?*, edited by B. Carr. Cambridge: Cambridge University Press.

———. 2007b. "Entropy, Emergence, and the Physical Roots of Natural Evil." In *Physics and Cosmology: Scientific Perspectives on the Problem of Natural Evil*, edited by N. Murphy, R. J. Russell, and W. R Stoeger. Vatican City State and Berkeley, CA: Vatican Observatory Publications and the Center for Theology and the Natural Sciences.

———. 2007c. "Reductionism and Emergence: Implications for the Interaction of Theology with the Natural Sciences." In *Evolution and Emergence: Systems, Organisms, Person*, edited by N. Murphy and W. R. Stoeger. Oxford: Oxford University Press.

———. 2011. "Emergence, Directionality and Finality in an Evolutionary Universe." In *Biological Evolution: Facts and Theories—A Critical Appraisal 150 Years after "The 1,"* edited by G. Auletta, M. Le Clerc, and R. A. Martinez. Rome: Gregorian and Biblical Press.

Van Till, H. J. 1986. *The Fourth Day: What the Bible and the Heavens Are Telling Us about Creation*. Grand Rapids, MI: Eerdmans.

CHAPTER FOUR

Beyond Horatio's Philosophy

Biological Evolution and the "Plurality of Worlds" Concept

Martinez J. Hewlett
DEPARTMENT OF MOLECULAR AND CELLULAR BIOLOGY, UNIVERSITY OF ARIZONA AND DOMINICAN SCHOOL OF PHILOSOPHY AND THEOLOGY, GRADUATE THEOLOGICAL UNION

There are more things in heaven and earth, Horatio, than are dreamt of in your philosophy.

SHAKESPEARE, *HAMLET,* ACT 1, SCENE 5

The discourse in the Western world on the possibilities of life on other worlds is as ancient as the philosophical tradition itself. Even to the present day the latest missions to Mars, aimed at the discovery of water and molecular evidence for the possibilities of life, excite the popular imagination at the same time that they challenge our notion of self and our place within the cosmos. Hamlet's words to Horatio, after his first encounter with the ghost of his dead father, the king, ring true for us in preparation for our initial, and perhaps inevitable, encounter with the other.

The Ancient and Medieval Arguments

But just what is our philosophical preparation for this? Within the Western tradition the idea of worlds other than our own goes back at least to Greek intellectual reflection. In that setting, it was the atomists, led by

Democritus and Leucippus, who held that other worlds were not just possible but, in fact, inevitable. If, they maintained, the Universe is infinite, and if everything within it is made up of the same material particles as our own world, then it must be that other such worlds exist (Ruse 2001). Although what the atomists meant by "matter" was very different from our current scientific interpretation of that word, their position with respect to the existence of other worlds is in many ways echoed by our modern enthusiasts who search for signs of life. I would argue that this philosophical stance of inevitability is implicit in the Search for Extraterrestrial Intelligence (SETI) Project.

Countering the atomists on this topic was the Athenian school founded by Socrates and continued by his two disciples Plato and Aristotle. While differing in their approach to understanding the observable world, they were united in their rejection of the possibility of other worlds.

There were two objections. First, both Plato and Aristotle argued that the Creator of all that can be observed is unique and therefore there can be only one such creation. Other worlds, in this case, would represent other creations, which would not be possible. A second objection involved an understanding of physics and the structure of the cosmos as Aristotle conceived them to be (Aristotle 1952).

In his treatise *De Caelo* (*On the Heavens*) Aristotle makes the case for a Universe with the Earth at the center and the heavenly bodies circling in ever-increasing spheres. The elements that make up the Earth were fire, air, water, and earth, whereas the heavens consisted of the ether or quintessence, the "fifth element." Given this geocentric view, Aristotle concluded that bodies like our own and the things around us had a "natural place." Rocks, for instance, fell to the ground when dropped. How could it be that rocks existed elsewhere in the cosmos, when their natural tendency would be to fall to the Earth? Therefore, other worlds that would necessarily consist of the same four elements could not exist beyond the Earth (Aristotle 1952).

The Platonic tradition and subsequently the Aristotelian approach became the basis for the intellectual life of Christian scholars in the West. St. Augustine used Platonic idealism to formulate his support for Christian dogma. St. Thomas Aquinas incorporated the newly reencountered philosophical methodology of Aristotle to understand the nature of creation and humans in a Christian philosophical and theological context.

Throughout this period of Western history the cosmological model did not change. For both Aquinas and Aristotle, the geocentric concept worked. It must be emphasized that this model did not incorporate our modern understanding of the nature of the Moon, the planets, and the stars. In fact, for both of these philosophers the Earth itself consisted of the corruptible four elements—fire, air, earth, and water. Everything else, the stars and planets, existed in the realm of the incorruptible fifth element, the quintessence.

When Aquinas questioned the possibility of extraterrestrial life, he did so within the framework of this Aristotelian worldview. Given that there is only one Universe, if other worlds did, in fact, exist, they would have to consist only of quintessence. Therefore, beings on such worlds could only be of the same substance. Could the planets and stars be populated by beings like us? His answer, in true Dominican style, was both yes and no. To follow his argument, it is necessary to see what St. Thomas meant by "like us." If he assumed that this meant "consisting of the same material substance," then the answer is no. Because there is only quintessence to work with, there can be no bodies such as we know them, no organic life, in modern terms. If, on the other hand, by "like us" he meant rational intelligences, then the answer is a qualified yes. Aquinas contended that rational intelligence does not necessarily require a physical body. After all, the angels are a form of intelligence, he argued. Therefore, beings could exist on the planets and stars, but they would not be like us physically (George 2001).

This is an interesting philosophical position. After all, a complete astrobiological consideration today includes the possibility of life-forms that are not carbon-based—and those that, while still carbon-based, rely on different nucleic acid bases and amino acids than terrestrial life does. This notion is not restricted to the realm of science fiction writers. For instance, serious consideration has been given to the possibility of silicon-based life-forms (Cairns-Smith et al. 1992).

The Critical Break

Any conclusions about whether life exists elsewhere than on Earth and what form it might take are tied inevitably to what those other worlds might be like. And so, with the Copernican revolution comes a distinct shift in this astrobiological discussion.

The geocentric cosmological model of Claudius Ptolemaeus (Ptolemy), published in about 150 CE, elaborated on the Aristotelian scheme. Ptolemy's model, however, was in essence a calculating tool. Using this geocentric arrangement, he could accurately predict the positions of celestial objects over the course of time. It naturally followed that a physical model of the Universe would be developed that mimicked the predictive model. Thus, for more than a thousand years Western culture considered the Earth to be the center of the Universe.

Over time and with more precise observations, the use of the Ptolemaic calculating tool became more complex. To explain finer movements, such as retrograde motion, the apparent reversal in direction of movement seen for the planets, it was necessary to apply epicycles to the calculating tool. Such ad hoc corrections were becoming more cumbersome and making the tool less likely to provide accurate predictive information in the long run.

Nicolaus Copernicus tackled this problem and came up with a revolutionary solution: heliocentrism. Others before him, including certain ancient Greek and medieval Muslim commentators, had suggested this as a model. In fact, it has been argued that Copernicus relied on these sources in constructing his calculating model. Nevertheless, it is the Copernican revolution to which we point as a pivotal development in the history of Western science. Certainly this paradigm shift had a greater effect on our view of the physical Universe than on our view of its biological features, but the way in which we envision living things is dependent upon the model we have of the cosmos. Nonetheless, a detailed model of living systems and their origin remained beyond the purview of science until the Darwinian revolution, which Francisco Ayala contends "completed the Copernican revolution" (Ayala 1998).

To hold that the Earth, along with the other planets, moved in an orbit around the Sun was entirely contrary to the cosmological view of Western culture, including both theological and philosophical positions. As a result the Copernican model met with resistance, first within the burgeoning Reformation community and later within the Roman Catholic Church.

Although one might like to think that the problem lay totally within a strict interpretation of biblical texts regarding the movement of the Sun, the actual philosophical issue was much different. In the Aristotelian/Ptolemaic cosmology, the planets and stars consisted of the incorruptible quintessence, not of the corruptible elements of fire, air, earth, and water. To elevate the Earth to the position of the planets seemed presumptuous at best (Brooke 1998). However, Old Testament cosmology, which was not even Aristotelian, did seem to conflict with the heliocentric hypothesis of Copernicus. Thus, some of the conflict took on theological dimensions that verged on what some judged to be heresy.

Among the supporters of Copernican heliocentrism was a former Dominican friar, Giordano Bruno. The Campo di Fiori in Rome has, at its center, a statue commemorating Bruno, because this was the site of his execution by burning at the stake on February 17, 1600. Often cited as the first "martyr" of the scientific age, his death at the hands of Church authorities was officially for heresy. However, most scholars agree that the heresy in this case was not his support of Copernican models but instead a whole series of issues that contradicted Catholic theology. Only one of these charges related to his astronomical opinions. Interestingly, it was about the plurality of worlds and their potential eternity, not the heliocentric model (Crowe 1999).[1]

The Copernican model was persuasive, and soon many scientists within and outside of astronomy had taken its conclusions on board in their thinking. Evidence began accumulating in support of the model. Galileo's observations of the moons of Jupiter orbiting that planet gave visual, if only indirect, support to the new model. But not all astronomers were convinced. Others, including Tycho Brahe, did not fully accept this system. Brahe proposed a combined model, a geoheliocentric system, in which the Sun with the orbiting planets revolved around the Earth.

Eventually the weight of repeated observations pushed the acceptance of the Copernican system. René Descartes, who was considered heliocentric in his view, tried to imagine a way in which the Earth could still remain effectively stationary in a vortex of motion. By this time, however, most scientists were firmly committed to the new cosmology. In many ways this move represented the break between science and the hegemony of the Church, although the extent of this rift has been overemphasized in much of modern literature.

The new cosmological model addressed only issues of the physical arrangement of the Solar System and, by implication, the Universe. What about the nature of life-forms, if any, present on extraterrestrial bodies? At that time in the history of science no credible natural account existed. As a result, the default position was special creation, by which is meant the view that the Universe and everything in it was created in their present form by divine action.

Opinions varied about the possibility of other worlds with life. Descartes was skeptical, although he did allow that if animals or people were present on the Moon, telescopes might one day have sufficient resolution to see them.[2] One of the most elaborate descriptions was that of Bernard le Bovier de Fontenelle, the French poet and commentator on philosophy and science. His book *Conversations on the Plurality of Worlds*, published in 1686, defended the Copernican model and proposed that all of the fixed stars had planetary systems, replete with inhabitants similar to those on Earth. The book was a landmark for its time, because it was written in French rather than Latin and because one of its central characters was a woman who was expert in scientific matters.[3] Fontenelle's scheme was finally taken up by a reputable astronomer, Christiaan Huygens. *Cosmotheoros*, published in 1698 shortly after Huygens's death, made the case for extraterrestrial life along the same lines as Fontenelle, although Huygens's approach was more scientific.[4]

The Last Great Debate on the Plurality of Worlds: Whewell and Brewster

By the beginning of the nineteenth century, Western science had advanced on a number of fronts, particularly in the physical sciences. Newtonian mechanics and optics were fully developed fields. Boyle had laid the foundation for understanding thermodynamics, and electricity and magnetism were active areas of investigation for Coulomb and Gauss. In astronomy, with the continual improvement of observational instruments, new discoveries were being made at a rapid pace. Herschel had identified the planet Uranus, Laplace had developed a rigorous description of celestial mechanics, and Halley was charting the movements of stars and comets. The understanding of the age of the Earth had changed dramatically with the proposal of deep geological time by Hutton.

Biology had also entered the beginning of its revolutionary cycle with the establishment of the hierarchical classification system of Carle Linné near the end of the eighteenth century. The true paradigm shift would come in 1859 with the publication of Darwin's *The Origin of Species*. However, in 1844 Robert Chambers, a Scottish journalist, anonymously published a book entitled *Vestiges of the Natural History of Creation*. In this work he proposed transmutation or evolution as the means by which everything we observe came to be, whether it be stars and nebulae, the geology of the deep-time Earth, or the living world. The book caused quite a stir in Britain, and reactions were widespread. In fact, Tennyson's famous "nature red in tooth and claw" and other evolution references in his great poem *In Memoriam* were written before Darwin's publication and in reaction to reading *Vestiges*. In *The Origin of Species*, Darwin mentions that Chambers's work has "done excellent service in this country in calling attention to the subject, in removing prejudice, and in thus preparing the ground for the reception of analogous views," but he criticizes it for providing no valid scientific evidence or mechanism (Darwin 1952).

Into this mix in the nineteenth century arrived two protagonists in the plurality-of-worlds debate. William Whewell was a Cambridge scientist, philosopher, theologian, and Anglican priest. It can be argued that he was one of the greatest influences on philosophy of science in the nineteenth century. He coined the term "consilience," used most recently by E. O. Wilson, as well as the term "hypothetico-deductive method." He championed the use of inductive reasoning as the hallmark of good scientific investigation. The other player in this debate was David Brewster, a Scottish scientist and Presbyterian layman.

Whewell initially was in favor of the idea of other planets with lifeforms. In 1830 he was giving sermons extolling the wonders of these possibilities. However, in 1853, perhaps in reaction to Chambers's *Vestiges*, he published *Of the Plurality of Worlds*, a semi-anonymous essay attacking the very idea that there could be life, especially intelligent life, anywhere except on planet Earth. His main issues were (1) the geological record showing that the length of humankind's presence on Earth represents a minute fraction of the deep-time age of the Earth, (2) the uninhabitability of the other planets in the Solar System, (3) the lack of any observable planets (at that time) anywhere else in the Universe, and (4) various philosophical and theological issues stemming from his ideas about design and lawlike behavior.

Brewster countered in an 1856 publication entitled *More Worlds Than One: The Creed of the Philosopher and the Hope of the Christian*. His objections were focused on two main points: (1) God's greatness was revealed by His unlimited creativity, and (2) humans were, in fact, too trivial to be the highest and only intelligent result of this creation.

Michael Ruse, one of the best-known contemporary philosophers of science, has called Whewell's publication and the ensuing debate "the biggest scandal between Chambers's *Vestiges* of 1844 and Darwin's *Origin of Species* of 1859" (Ruse 2001). Indeed, the debate raged back and forth for a number of years. Notice that in each case the proponents were taking a distinctly theological position, because the only plausible explanation for the presence of living things was a divine special creation.

This debate, I believe, represented the last time in Western science and culture that these positions could be reasonably taken. With the publication of *The Origin of Species*, explanations concerning the beginning and evolution of living forms on our planet took on a decidedly scientific direction.

Scientific Explanations for the Origin of Life

Darwin's model for biological evolution does not consider how life originated. In fact, Darwin says very little about this, directing his attention almost exclusively to the problem of how the diversity of living forms might have arisen. At one point, in correspondence, he did speculate on the possibilities for the origin of life:

> It is often said that all the conditions for the first production of a living organism are now present, which could ever have been present. But if (and oh! what a big if!) we could conceive in some warm little pond, with all sorts of ammonia and phosphoric salts, light, heat, electricity, &c., present, that a proteine compound was chemically formed ready to undergo still more complex changes, at the present day such matter would be instantly devoured or absorbed, which would not have been the case before living creatures were formed.[5]

Nevertheless, with the publication and final scientific acceptance of his model, the discourse naturally moved to questions about how life might have gotten started in the first place.

Darwin's theory supposed that naturally occurring variations in species would lead to reproductive advantages for certain variants under specific environmental selection conditions. As a result of these advantages the descendants of individuals expressing these variations would tend to become more represented in the population, especially when the available resources for reproduction were limited. This "survival of the fittest," as Herbert Spencer came to name it, predicted that present-day species were descendants of these contingent pathways.

Using these same concepts, early workers began to model how life itself may have arisen under prebiotic conditions present on the young planet. The speculation was that molecules necessary for the construction of the living forms that we observe might have arisen in a more or less "random" fashion. The word "random" here is not meant in the strictly mathematical sense of "all possible variations" or "no discernible pattern." Rather, it implies that chance interactions under these conditions might lead to one or more types of molecules that have properties that ultimately lead to self-replicating systems. The reproductive advantage of such systems under certain conditions would then provide the selectability inherent in the Darwinian model.

In fact, the earliest considerations of just what these initial conditions might be reflected Darwin's "warm little pond" scenario. We can see this in the speculations of Aleksandr Oparin and independently those of J. B. S. Haldane. Into the warm little pond, which could be the prebiotic sea of the early Earth, they imagined an energy source such as lightning entering to drive the synthesis of the potential organic molecules. It was Haldane who coined the expression "primordial soup" that has come to be the sound bite of abiogenesis, the search for scientific models of life's origins (Oparin 1953).

Finally, Harold Urey and Stanley Miller in 1953 "did the experiment," as we like to say in science. Their now-famous apparatus attempted to mimic conditions that were theoretically present. The sealed flask contained water, ammonia, hydrogen, and methane, and instead of lightning they introduced an electric spark. Sure enough, various kinds of complex organic molecules could be found in the resulting mixture, including amino acids, sugars, and lipids (Johnson et al. 2008).

So compelling was this and similar experiments that the energy-driven synthesis of complex organic molecules which are building blocks of primordial life has become a tacit assumption in the field of astrobiology. In 1961, Frank Drake suggested his well-known equation that estimates the potential number N of communicating civilizations we might encounter in a search for extraterrestrial intelligence: $N = R^{*} \times f_p \times n_e \times f_l \times f_i \times f_c \times L$.

The terms in the Drake equation express the probability for various conditions to be true.[6] Notably, the factor f_l is the likelihood that if certain conditions are present on a planet, life will develop. Drake's original estimate for this term is $f_l = 1$. That is, given the starting conditions deemed necessary, 100 percent of the time life will develop. Notice that this conclusion is based on an observed sample size of one.

Putting aside for now discussion of this estimate, it is interesting to note how the plurality-of-worlds debate has changed over time. As a result of the Copernican paradigm shift, it became possible to envision planetary systems other than our own. However, it is due to the Darwinian paradigm shift that the presence of life on such extraterrestrial planets could be thought of as the result of physical and chemical processes like those—or similar to those—leading to life on Earth.

Horatio's Philosophy Revisited

In Shakespeare's *Hamlet*, the Prince and his friends, including Horatio, are students at the University of Wittenberg. Horatio is presented to us as a model of rational thought, a person steeped in the disciplines of natural science and logic. In Hamlet's story about his encounter with the ghost, Horatio is confronted by something for which he has no explanation within his system of understanding, his cosmological view.

In many ways astrobiology puts us into a dilemma like Horatio's. How are we to think about the presence of life elsewhere in the cosmos? Most importantly, what are we to make of possible contact with extraterrestrial intelligence (ETI)? How does this challenge the concept of who we are and why we are here? What should be our attitudes toward extraterrestrial life (ETL) and ETI? How should we treat it?

Much of this issue surrounds how we define ourselves. What does it mean to be human? Following on this we can also ask: what does it mean to be humanlike?

Here again the Darwinian model colors the discussion. Placing ourselves within an evolutionary continuum of species on our planet, we might define ourselves by biological criteria, including the fine structure of our genome.

Clearly, the discovery of ETL would undermine the uniqueness of our biological story. Furthermore, solid evidence for ETI would confirm that technological sophistication as a product of a rational mind can indeed occur as an outcome of an evolutionary sequence. However, these physical and biological discoveries, as astounding to us as they would be, would not in and of themselves be exhaustive answers to the questions above.

Essential to any consideration of life and intelligent life here or elsewhere in the Universe—of who we are and how we are to interact with and treat extraterrestrial life and extraterrestrial intelligence—is what David Chalmers (1995) calls "the hard problem," the mind/brain relationship that leads to the subjective experience we call consciousness. A completely physicalist explanation, one that assumes the ontological position that material reality is all that exists, equates mind or consciousness with brain states. Using this assumption, imaging studies with functional MRI or quantitative EEG purport to "demonstrate" the physiological activity of certain brain areas in correlation with specific mental tasks. In this model, consciousness is the result of the computational complexity of the neuronal networks and their chemistry (Jeeves 1998). For the physicalist, then, such a model would explain ETI as a remarkable yet somewhat inevitable outcome of the processes leading to ETL.

Why is this an incomplete explanation? It is clear from the nature of consciousness itself, especially from the subjective experience of conscious entities, that this physiological correlation does not encompass certain features. For instance, such strictly physicalist models do not explain our inner subjective experience of qualia or of time flow, nor our ability to exercise choice in what we call free will. A classical computational model cannot account rigorously for these features of our everyday understanding of what it means to be conscious. A more complete treatment of these arguments can be found in Michael Lockwood's book *Mind, Brain, and the Quantum* (Lockwood 1989). If our theory of mind—that is, our imputation of mental states in others—is enlarged to include ETI, then physicalism also becomes an incomplete explanation for this expanded set of conscious rational beings.

What model shall we use? I would like to offer two possibilities, seemingly unrelated. First, consider St. Thomas Aquinas's take on what it means to be human. His argument was that the human soul, or consciousness, if you will, is that which makes us be what we are. For Aquinas, the soul is defined as the "first principle of being," or, in philosophical terms, the formal cause of what it is to be a human person (Hibbs 1998).

The second approach to this complex issue is to consider that mind or consciousness is the product of the quantum properties of the brain. In this model, proposed by Stuart Hameroff and Roger Penrose, quantum states of brain components, such as the alpha and beta tubulin subunits that make up the neuronal microtubules, effectively produce a quantum computational state. Properties of this state include both quantum superposition and quantum entanglement (Hameroff and Penrose 1996). In their view, the model accounts for the features of consciousness that cannot be dealt with using classical neuronal biochemistry and network analysis.

What do a medieval philosophical position and a modern quantum theoretical model have in common? My answer has to do with the medieval and modern notions of matter. For Aquinas, in the Aristotelian tradition, matter was potential or material cause and could be actualized only by a formal cause. In the modern understanding of the nature of physical reality, the idea of something coming into actual existence from a kind of probability is inherent in the quantum model, at least for the standard interpretation. Thus, the idea of the soul as the first principle of life, the formal cause of the living being, and the quantum computational model have in common, among other things, the notion of probability leading to actuality.

I do not mean to suggest that Aquinas was a quantum theorist or that modern physicists are Thomistic. Instead I would argue that the philosophical concept of formal cause, which in more contemporary terms is nothing other than the network of dynamic organizational relationships that make something what it is, might be recovered in a way that has explanatory power. Because the natural world, in the Aristotelian/Thomistic view, is all about material cause being actualized (form acting on matter), science is already examining the consequences of formal causation. A consideration of the human person as the result of formal causation might include both physical (for instance, systems biology and emergent properties) as well as nonphysical aspects. The quantum mechanical level of understanding may provide a critical link between these two apparently disparate points of view.

What then of the questions I posed at the beginning of this section? How will having a more complete model of life in general, and human life with a rational mind in particular, lead us to answers?

First, the discovery of ETL will add an interesting data set in support of our current models of biogenesis. We will be able to conclude that, if the conditions are right, life can occur in more cases than just the one here on Earth. The chemical basis of ETL would speak to questions about the range and variety of ingredients and conditions necessary for life.

However, an encounter with ETI and the realization that rational minds exists outside of our own species (*pace* current work with certain animals on this planet), pushes us well past a mere justification of our scientific models. This has the potential to bring philosophical, ethical, and theological issues all to the table, especially if our considerations include more than physicalist explanations.

Ted Peters, in a recent paper (Peters 2011) and elsewhere in this volume, has explored these implications, with an emphasis on their ethical and religious dimensions. Using results from an online survey tool as well as interviews with theologians and scientists, Peters concludes that knowledge of ETL and ETI would expand and enrich, rather than threaten, ethical and religious viewpoints.

I contend that Aquinas would have agreed with Peters in this case, both theologically and philosophically. From a Thomistic standpoint, this encounter could only enrich the definition of what it means to be a human person by enlarging the category "human" to include all beings with a physical body and a rational soul, however that body appears to be constituted.

With astrobiology in general, and the SETI Project in particular, we have an opportunity to think more deeply about these issues and formulate our response. When we do meet the other, how will we behave? When we sit across the table from ETI, will we know that we are related?

Notes

1. For two different views of Bruno's condemnation, see the website for the Galileo Project (http://galileo.rice.edu/chr/bruno.html, last accessed May 14, 2009) and the Web version of the Catholic Encyclopedia (www.newadvent.org/cathen/03016a.htm, last accessed May 24, 2009).

2. Comments regarding the possibility of animals or people on the Moon can be found in correspondence and journal entries during Descartes's time in the Netherlands.

3. The full text of *Conversations on the Plurality of Worlds* can be found at Google Books (http://books.google.com/books?id=VGoFAAAAQAAJ&pg=PA1&dq=Fontenelle+Bernard+inauthor:Fontenelle#PPP24,M2, last accessed on May 26, 2009). A dated but still relevant discussion of Fontenelle's place within the French Academy during the Enlightenment can be read in Leonard Marsak, "Bernard de Fontenelle: The Idea of Science in the French Enlightenment," *Transactions of the American Philosophical Society* 49 (1959): 1–64.

4. A Web version of Christiaan Huygens's *Cosmotheoros*, in English translation, can be found at www.phys.uu.nl/~huygens/cosmotheoros_en.htm (last accessed on May 26, 2009). A link is also provided to the original Latin edition, as well as to a French translation.

5. Charles Darwin, in a letter to Joseph Hooker, February 1, 1871. The letter is referenced in the Darwin Correspondence Project (www.darwinproject.ac.uk/darwinletters/calendar/entry-7471.html, last accessed on May 24, 2009), but the complete text is not available electronically. The relevant passage can be found in the Internet Encyclopedia of Science, www.daviddarling.info/encyclopedia/D/DarwinC.html (last accessed on May 24, 2009).

6. The terms in the Drake equation are as follows: N = the number of civilizations that might communicate with us in our Galaxy; R^* = the yearly rate of star formation in our Galaxy; f_p = fraction of stars with planets; n_e = average number planets per star with the potential to support life; f_l = fraction of those planets that do develop life, f_i = fraction of those planets that develop intelligent life; f_c = fraction of those planets that become technologically sophisticated and can generate communication signals; and L = the length of time that those signals are being sent. A good, brief discussion of the Drake equation can be found at the SETI League site, www.setileague.org/general/drake.htm (last accessed on May 24, 2009). An interactive version of the equation can be found at www.activemind.com/Mysterious/Topics/SETI/drake_equation.html (last accessed on May 24, 2009).

References

Aristotle. 1952. *On the Heavens (De Caelo)*. Translated by J. L. Stock. Chicago: University of Chicago Press.

Ayala, F. J. 1998. "Darwin's Devolution: Design without Designer." In *Evolution and Molecular Biology: Scientific Perspectives on Divine Action*, edited by R. J. Russell, W. R. Stoeger, and F. J. Ayala. Vatican City State: Vatican Observatory Foundation.

Brooke, J. H. 1998. "Science and Religion: Lessons from History?" *Science* 282: 1985–1986.

Cairns-Smith, A. G., A. Hall, and M. Russell. 1992. "Mineral Theories of the Origin of Life and an Iron Sulfide Example." *Origins of Life and Evolution of the Biosphere* 22: 161–180.

Chalmers, D. 1995. "Facing Up to the Problem of Consciousness." *Journal of Consciousness Studies* 2: 200–219.

Crowe, M. J. 1999. *The Extraterrestrial Life Debate, 1750–1900*. Mineola, NY: Dover.

Darwin, C. 1952. *The Origin of Species*. 6th ed. Chicago: University of Chicago Press.

George, M. 2001. "Aquinas on Intelligent Extra-Terrestrial Life." *Thomist* 65: 239–258.

Hameroff, S., and R. Penrose. 1996. "Conscious Events as Orchestrated Space-Time Selections." *Journal of Consciousness Studies* 3: 36–53.

Hibbs, T., ed. 1998. *Aquinas: On Human Nature*. Indianapolis: Hackett.

Jeeves, M. 1998. "Brain, Mind, and Behavior." In *What Ever Happened to the Soul?*, edited by W. Brown, N. Murphey, and H. N. Malony. Kitchener, ONT: Fortress Press.

Johnson, A. P., et al. 2008. "The Miller Volcanic Spark Discharge Experiment," *Science* 322: 404.

Lockwood, M. 1989. *Mind, Brain, and the Quantum*. Cambridge, MA: Blackwell.

Oparin, A. 1953. *The Origin of Life*. Translated by Sergius Morgulis. New York: Dover. Originally published in 1924 in Russian.

Peters, T. 2011. "The Implications of the Discovery of Extra-Terrestrial Life for Religion." *Philosophical Transactions of the Royal Academy* 369: 644–655.

Rodis-Lewis, G. 1998. *Descartes: His Life and Thought*. Ithaca, NY: Cornell University Press.

Ruse, M., ed. 2001. *Of the Plurality of Worlds, by William Whewell*. Chicago: University of Chicago Press.

CHAPTER FIVE

The Wonder Called Cosmic Oneness

Toward Astroethics from Hindu and Buddhist Wisdom and Worldviews

Nishant Alphonse Irudayadason
JNANA-DEEPA VIDYAPEETH: PONTIFICAL INSTITUTE OF PHILOSOPHY AND RELIGIONS

In this chapter I attempt to show how the Hindu and Buddhist worldviews affirm universal harmony and how this can be paradigmatic for evolving astroethics. We shall explore the possible connections between key insights and intuitions of Buddhist and Hindu thought and our rapidly expanding understanding of the Universe, including our knowledge of life and conscious life here on Earth and possibly elsewhere. In particular, what Buddhist and Hindu perspectives and convictions are consonant with the broad findings of cosmology, astronomy, biology, and astrobiology? And what basis do they provide for ethical attitudes and guidelines for our interaction with environments, organisms, and societies here on Earth and elsewhere in the cosmos?

The Universe forms an organic and indivisible whole. All its parts are interdependent. The whole becomes possible only in the collaboration of all its parts. This is one of the important insights of the great philosophical tradition since Plato and finds echo in the creative thinking of Aristotle and Plotinus, in the discourse of the Stoics and the Neoplatonists, and later in the treatises of Spinoza and Hegel. In the words of Emperor Marcus Aurelius, "All things are woven together and the common bond is sacred, and scarcely one thing is foreign to another" (Aurelius 1998). All beings are intertwined in the whole like the threads of a single fabric. A life breath unites them all, and they participate in and contribute to the universal harmony.

Thus this idea of cosmic oneness is not exclusive to Hindu and Buddhist philosophies. In fact, a stream of Western philosophy, along with modern science, and psychology promotes this view. The Stoics and the Neoplatonists have argued that the Universe possesses a single soul, in which all individual souls participate and are related to one another. Spinoza affirms, "We may easily consider the whole of nature to be one individual, whose parts, that is to say, all bodies, differ in infinite ways without any change of the whole individual" (Spinoza 2009/1673). His contemporary, the champion of monism in Western philosophy, categorically states: "Reality cannot be found except in one single source, because of the interconnection of all things with one another" (Leibniz 1965/1670). In this vision of the Universe, each singular point is a knot of relationships. Every particular being, writes Hegel, exists only in its relation to others: everything is connected and interdependent. Things are different, but they are not separated, and each carries the mark of all. The microcosm therefore reproduces the macrocosm.

Both physical sciences and human sciences have profoundly changed the course of the twentieth century, allowing mechanistic and deterministic models to move toward holistic approaches. What characterizes this new perspective is the idea that each object (an atom, a cell, an organism, a human being, a society)—forming an integrated whole through the dynamic relations of all its elements—is itself part of a further larger system. From this perspective, interdependencies and interrelationships are fundamental. In modern physics, the vision of the Universe as organic and dynamic is becoming more widespread. Elementary particles are also understood as interactions or interrelations rather than objects or substances. Theories of relativity and quantum mechanics have somehow paved the way for a new holistic approach.

Thus, advancements in modern physics have led to an understanding of the Universe as organic and dynamic. Physicists like David Bohm and Fritjof Capra affirm that, in some sense, the Universe is not only a living organism but also intelligent. They speak of the inseparability of the particles, forming a single cosmic fabric. Again, the Universe appears as an indivisible whole, all parts interconnected with each other and related to the whole. Each elementary particle appears essentially as a knot of relationships and events. It is interwoven with the other elements and exists only in interdependence within the cosmic whole.[1] Isn't this similar to the

Soul of the Universe of the ancient Greeks or the Great Atman of the Hindu-Buddhist traditions?

This interrelatedness is exemplified in the physical interactions of particles. For instance, a proton can become a neutron and a neutron can change into a proton, thus giving rise to photons (light). Both light and matter can transform into each other. Mass itself is energy, and this energy that constitutes the whole Universe can take any form. In short, modern physics seems to suggest that in the Universe, the game of interconnection is essential. This game is the unique dance of all the elements, and no particle is completely isolated. The Universe is a dynamic whole held together by the major binding forces: the strong and the weak, the electromagnetic and the gravitational nuclear forces. They energize and orchestrate the dance of the binding energies, creating nuclei, atoms, molecules, cells, organisms, planets, stars, and galaxies. Thus, everything in the Universe is interdependent, and the cooperation of all the elements seems to be a fundamental law, yielding novelty and complexity on the one hand, and harmony and unity on the other. This is undoubtedly an important aspect of what the physicist Trinh Xuan Thuan calls "the secret melody" in the heart of the Universe (Thuan 2005).

The human person is a part of this Universe. Her ancestors are the tiny particles of matter. Hubert Reeves calls her "stardust." She is ultimately the result of cooperation of the elements in the Universe. She participates in the cosmic dance and sings in her own way, the secret melody of the Universe. This is the approach adopted by the great biologist Gregory Bateson. He speaks of a structure that connects all living beings. He evokes the unity of biosphere and humanity and emphasizes the beauty of this ordered dynamic whole. In his deeply spiritual vision, beings are defined by their relations to other beings. Similarly, human persons as individual animated beings are part of the Universe.

As already mentioned, developments in psychology also have moved toward a holistic approach. Jung, for example, spoke of the collective psyche or the collective unconscious. Transcendental psychology, which studies transpersonal experience beyond individual limits, shows that no individual can be separated from others or from the environment. Though psychologists often tend to limit the scope of their research to social environments, it is not difficult to show that the social environment itself is part of a larger Universe, which has a place not only for humans but for other beings as well.

The emerging idea of holistic health also acknowledges the integrity of the human organism, as well as its deep connections with the greater cosmic whole. Illness is seen as a rupture of harmony, and restoration of health results, at least in part, from restoring its harmony with the Universe. This new approach of the physical and human sciences converges with ancient philosophical and spiritual wisdom: to be human is not to be solitary but to be in solidarity with all other beings. The human person belongs to the cosmos and is related to every other being in the cosmos.

The oriental spiritual traditions also affirm that human persons are united with all other beings. In Hinduism the idea of isolated existence is an illusion (*māyā*). "You belong to all and all belong to you," says Buddhism. All the waves are of the same infinite ocean, the unique conscious energy and love, which is Brahman in Hinduism, or the one who lives in the great void without forms in Buddhism. The Buddhist bodhisattva and the liberated Hindu are both beings inhabited by "the law of love." The bodhisattva who has reached the threshold of the Great Void (*nirvāṇa*) comes back to humans with "the blazing fire of compassion," and the Hindu liberated while living (*jīvan-mukti*) free from all selfish desires participates in the cosmic game of love.

This game unites all beings. The whole Universe is in a stone or in a star, and the stone and the star are in the human person. Each stone and each star is connected to the rest of the Universe, and every human person is bound to the stone and the star. "The soul of man is the whole Universe," says a Hindu sage. "Everything is in us," says Meister Eckhart. The human person is not an island lost in the vast cosmic ocean. She is called to design a huge project for achieving universal harmony. This is where ethics in a cosmic dimension can become meaningful. "Though One, Brahman is the cause of the many. Brahman is the unborn (*aja*) in whom all existing things abide. The One manifests [itself] as the many, the formless putting on forms" (*Ṛg-veda*). Religion should have a cosmic dimension. "If there is any religion that could cope with modern scientific needs it would be Buddhism," says Einstein.[2] This is because the Buddhist philosophy of science "takes the oppressive solidity out of the material and strips the spiritual of all its metaphysics, bringing it back to the immediate here and now" (du Pre 1984). If modern science can discover and learn about extraterrestrial beings, both Hinduism and Buddhism may help us to evolve our ethical thinking vis-à-vis such beings.

Here I shall attempt to show how the Buddhist and Hindu worldviews offer us philosophical underpinnings to evolve astroethics. Though both Buddhism and Hinduism have many diverse approaches to reality, I shall focus on the principle of nonduality. To do so, I shall first try to show how

Buddhist philosophy "deconstructs" the binary opposites of good and evil—paradoxically this binary is accepted as almost inevitable for any meaningful discourse on ethics—precisely to widen the scope of ethics beyond the web of human relations. Then I shall interpret one ancient Hindu text from the *Ṛg-veda*. Our hermeneutical reading of this text will be supported by other ancient narrative texts from Vedas, Upaniṣads, and the two great Hindu epics. In very symbolic mythical language, this ancient text affirms the principle of nonduality, thus pointing to a development of ethics on a more holistic and cosmic plane.

Buddhism

Traditionally, ethics presupposes the binary right/wrong, which has its metaphysical counterpart in the binary good/evil. Evil as such is not treated as a problem in the teachings of Buddha. This is primarily because in Buddhism, good does not possess the absolute value it has in the strict monotheistic religions. Buddhism does not evoke the idea of divine creation, paragon of good, within which evil can be understood as a radical and scandalous opposition to such a divine project. Moreover, Buddhism makes a distinction between conditioned (relative) and unconditioned (absolute) reality. Such being the case, there should be two kinds of "good" and two kinds of "evil." If Buddhism can speak at all of a "supreme good," it is the state of liberation or *nirvāṇa*, the cessation of suffering (*dukkha*), that is, of the conditioned state of *saṁsāra*, the endless cycle of births and deaths; this conditioned state is itself maintained by illusion, which can be said to be the "supreme evil." We must therefore consider two pairs of the "good/evil" binary: one in relation to the conditioned state, the other the opposition of the conditioned state understood as evil to the unconditioned state understood as good.

Buddhist Cosmogony

Before addressing the problem of good and evil, it is important to understand creation from the Buddhist perspective.[3] The Buddhist cosmogony is not so much meant to explain why the world is born as to describe how. We find one such narrative already from the very first text of the Pāli canon, the *Brahmajāla-sutta*. From the Buddhist perspective, like all phenomena, Universes come into existence one after another in an uninterrupted cycle of births and deaths, appearances and disappearances whose very first origin cannot be traced. At the birth of a Universe, in the abode of the Brahmās appears one Brahmā[4] according to his *karma*—the result of his good deeds in the previous birth. Endowed with feelings, this Brahmā soon feels unbearably lonely and wishes the rebirth of his companions, who in fact appear soon but on account of their own *karma*. But Brahmā thinks that he is the one who has awakened them and is convinced that he is the Mahābrahmā, the one who creates.

It becomes clear that there is no act of creation enacted by the Mahābrahmā. What happens independently of him is only the continuous cycle of births and deaths. In other words, Mahābrahmā believes himself to be the creator, a belief that is also shared by others who join his company. What is created in fact is the self, which is nothing but a mere illusion, a mere mental construction. It is not the world that is created by Mahābrahmā, but only "his" world. He becomes the builder of his "world." The Pāli canon of Buddhist texts uses the term "world" with a specific meaning. The "world" is a mental construction that every sentient being superimposes on reality, and this mental category is constructed from the illusion of the self. In this regard, the idea of good and evil does not fall within the purview of the world except through its relation to the self. It is a fundamental given, not of the "world as it is," but of "our world," which is purely a mental construction.

The greatest obstacle in our relation to others is precisely our constant temptation to impose our categories on them. We do not try to encounter the strange other in her otherness but the other of our mental construction. Emmanuel Lévinas, the champion of alterity in contemporary Western philosophy, challenges us to open ourselves up to meet the other in her otherness. Buddhism leads us a step ahead to identify the basis for the mental construction that prevents us from relating to the other as she is as the self and shatters the idea of the permanence of the self. This has two important implications for astroethics. First, by deconstructing the self, we are better prepared to meet other forms of life more meaningfully without reducing them to the categories of our mind, thus affirming their dignity in their own right. Second, if the concept of the self is an illusion, then the concept of "alien" is equally an illusion. This being the case, then, beyond all duality there exists a fundamental interrelatedness of all beings, which reminds us of our respectful coexistence.

The Idea of Dependent Co-Arising

This image of the builder and her house taken from a famous passage of two verses of the *Dhammapāda*, is supposed to express Buddha's exclamation immediately after his enlightenment:

I have coursed through the whirl
Of numerous lives
Seeking, but not finding,
The builder of the house.
Pain is birth over and over again.
Builder of the house—you are seen!
Never again will you build a house.
These rafters all broken,
The roof of the house destroyed,
The mind, free from the conditioned,
Has come to the end of cravings.

(*Dhammapāda*, verses 153–154)

This central idea of "building" is developed in the law of "dependent co-arising" (*paticcasamuppāda*),[5] which is deep, difficult to understand, hidden, excellent, beyond subtle, and wise reasoning accessible only through direct experience, as defined by Buddha himself shortly after his enlightenment. He was reluctant to reveal it to humanity, for those blinded by attraction and repugnance cannot understand this doctrine, which goes against ordinary thinking.

It is remarkable that in these proclamations of the enlightened Buddha there is no explicit mention of the concept of nonself (*Anātman*), which is usually presented as an essential part of Buddhist doctrine. The self in fact is none other than the builder sought in vain. What is revealed is instead the absence of the builder when the construction lets itself disappear and gives way to the reality. This is well summed up by Buddhaghosa in his famous *Visuddhimagga*: "There is suffering, but none who suffers; doing exists although there is no doer" (Buddhaghosa 2010). What exists in the domain of relative reality is only the dependent co-arising—the core of Buddhist teaching—to which the doctrine of no self is only a corollary, a logical consequence. This is because everything is dependent and nothing exists in itself. "The principle of dependence provides a philosophical basis for relating oneself to the natural world. It allows for the development of a feeling of kinship with nature" (Kalupahana 1995).

At the individual level, the self itself is a composition of five aggregates (*skandha*), arbitrarily designed as an independent and permanent whole through mental construction. Inasmuch as the builder disappears with the destruction of the building, when we look for the self beyond the five aggregates that compose it, it disappears with the decomposition of the five aggregates. Vasubandhu in his *Abhidarmakośabhāṣyam* gives the example of a string that is only the sum total of the strands that compose it, without any thread around which it is built (Vasubandhu 1990). At the origin of this mental construction lies an illusion. As long as this illusion persists, we will look for the builder in vain. When the illusion disappears, both the building and the builder disappear forever and this is the state of enlightenment. But for those who have not yet attained enlightenment, the illusion remains and every action is performed according to the illusory self. Just like the Mahābrahmā, who believes himself to be the creator of the phenomena produced around him, we live in the building with the conviction that we are the builder.

The idea of dependent co-arising affirms that ontologically we are dependent on others, including nonhumans, both for coming to be and for continued existence. Within this worldview, just as environmentalists insist that we need to care for nature to maintain ecological balance, we need to care for all forms of life, not only as a moral ideal of altruism, but also for our very existence. Thus, the idea of dependent co-arising offers a philosophical basis for a meaningful astroethical paradigm.

Buddhist Understanding of Good and Evil

It is only in relation to pleasure and pain, attraction and repulsion, that good and evil assume meaning. Good and evil can only be "my" good and "my" evil—good and evil always in relation to the self. "Going beyond good and evil was an idea common to all of India, including its Hindu and Buddhist traditions" (Hindery 2004). For the idea of good and evil to exist, it is imperative that there be an absolute around which the relation between the pair can be established. The paradox of dualism is precisely that the relation between the two requires an absolute, just as the two scales of a balance need a beam to maintain the balance. Only in reference to the self can good and evil exist. The problem lies not in the components of the relation but in the composition itself, not in the relative terms but in the relationship between them. This is the doctrine "against the current" that the Buddha taught. There is no problem of evil per se; there is instead a problem born of the "good/evil" relation.

The ancient Buddhist texts employ two pairs of terms to describe the good and evil. The first binary is from the everyday language: *pāpa* to designate evil and *Puñña* to designate good. There is also another binary, *kusala/akusala* (Payutto 1993).[6] The evil (*akusala*) is only the privation of the good (*kusala*). In fact, according to the Buddhist perspective, the "good/evil" binary is logically defined not in relation to any absolute but in comparison to a network of relationships. Evil (*akusala*) refers to the "three poisons" of the triad of attraction (*lobha*), repulsion (*dosa*), and illusion (*moha*), and evil arises not from one origin but from a multitude of conditions. And the good (*kusala*) is itself defined as the absence of attraction (*a-lobha*), of repulsion (*a-dosa*), and of illusion (*a-moha*). Here we find one of the main characteristics of the teachings of ancient Buddhism: the negative definition of the positive is paradigmatic in the very term denoting the "supreme good," *nirvāṇa* as absence of *dukkha*.

The definition of good and evil closes on itself in a vicious circle: evil is the absence of good, which is itself the absence of what characterizes evil and causes it. This vicious circle is precisely *saṁsāra*, the infinite cycle of births and deaths of the self, the eternal dependent co-arising. This binary "good/evil" is totally part of this *saṁsāra* subject to illusion and dualism. But the central problem in Buddhism, as we know, is not evil; it is *dukkha*, which is generally translated as "suffering." The Four Noble Truths are also generally summarized in the four formulas: the truth of suffering, the origin of suffering, the cessation of suffering, and the path to the cessation of suffering. *Dukkha* should be understood in opposition to *sukha*—well-being; *dukkha* is ill-being, "unhappy living," whereas *sukha* expresses purity, perfection, and fullness. The canonical definition of *dukkha* lends itself to three interpretations: the ordinary meaning of suffering (*dukkha dukkha*), suffering related to transience or the impermanence of phenomena (*vipariṇāma dukkha*), and the conditioned state of all phenomena (*saṅkhārā dukkhā*).

There are several levels of *dukkha*. First is the suffering related to the self—physical and psychological pain, disappointment, rejection by others as unworthy of love, and painful struggles in relationships. Second is suffering related to the phenomena that make a person doubt the actual existence of their real self, paving way to questions like "Who am I?" and "Am I?" Third, and more profoundly, is the fact that this is an illusion, that the self is just a mental construct and does not exist "in itself." An ordinary person faces *dukkha* in the first sense and perhaps at best in the second sense if she has some spiritual or existential concerns. Even to perceive the third aspect of *dukkha* requires the extraordinary ability to see reality as it is; it is perceived only by those who have already "entered the stream," having abandoned illusions.

The Buddhist teaching takes into account these successive realizations and adapts them accordingly to the capabilities of different audiences. To those who are immersed in total illusion, it proposes the way of what can be considered to be "the pursuit of the good as a lesser evil," the reduction or elimination of suffering in the first sense, as per the law of *karma*. To those whose illusions begin to crumble like a house that is on the verge of collapse, but who are still seeking the builder, it teaches the law of dependent co-arising. To those who have already "entered the stream" it teaches the ultimate emptiness of the self.

The concepts of good and evil are mostly found in the teachings addressed to neophytes, the first group of disciples, who are fully subject to the illusion and have not understood the value of renouncing the world either literally or figuratively. In the Pāli canon, the *Dhammapāda*[7] is undoubtedly the most emblematic text of this first type of teaching.[8] The main idea is that evil causes suffering, and it is better to do good in order not to suffer. Evil is clearly associated with the first meaning of suffering, whether it is moral or physical (*dukkha dukkha*). Suffering is evil. We suffer as much pain as we cause, because all actions bear fruit in conformity with their nature. Such is the law of *karma*.[9] At this first level of teaching within the framework of the law of *karma*, the objective is to reduce or eliminate suffering. Those who suffer have the single purpose of protecting themselves against suffering. Before considering the law of dependent co-arising, it is important not to suffer any longer from ordinary suffering, inasmuch as we cannot see through the muddy water and must therefore allow the mud to settle.

The ninth chapter of the *Dhammapāda*, entitled "The Evil" (*pāpavaggo*), speaks of evil in terms of the meaning of suffering in the second degree. Though this chapter offers only a few differences from the verses of the first chapter, where good and evil are presented according to the law of *karma*, the last two verses of this chapter bring in another perspective:

> Not in the sky,
> not in the depths of the sea,
> not by entering a cleft in the mountains,
> nowhere in the world can there be found
> a place where one might be released
> from detrimental actions.
> Not in the sky,
> Not in the depths of the sea,
> Not by entering a cleft in the mountains,

Nowhere in the world can there be found
A place where death would not prevail.

(*Dhammapāda*, verses 127–128)

After eight verses that oppose good to evil come these two verses that speak, not of a system of opposition, but of equivalence, in a dramatic style placing us in front of the inevitability of death. No one escapes the consequences of their actions, and no one escapes death. Even she who does well must die, because everything is impermanent. Here we are placed before the second meaning of *dukkha*—evil as the inevitability of impermanence and of perpetual transformation, and finally, the ultimate transformation of the self and death.

For those who move away from evil understood as ordinary suffering (*dukkha dukkha*), however well they may succeed in attaining the highest heavenly state, the illusion of the self would keep them in *saṁsāra* and leave them to grapple with the suffering linked to the impermanence. Even though Buddhist ethics advocates abstention from evil (*pāpa*) and doing good (*Puñña*), with the promise of a happy rebirth in the world of the gods, the last verse violently reminds us of the fact that the good of this world is only a "lesser evil," for it is of this world and hence always subject to the idea of self.

Even birth in the world of the gods can come as a major obstacle in the path of liberation. The idea of doing good, and thus benefiting from the good results of one's actions, participates fully in the perpetuation of the idea of the self. We may easily be convinced of the validity of our own actions and therefore of the very foundation of the idea of the self. Doing good can even become one of the worst dangers, as confirmed by the divine state where the satisfaction of pleasure is such that the sentient being is not even aware of the impermanence of phenomena and of his own ephemeral existence. This is why human life is precious. Human life is the only condition of existence that allows for a just measure of relativity between good and evil, suffering and pleasure, and for overcoming this dualism in order to be aware of the momentary nature of phenomena. Only a human being can enter into the flow. If Buddha himself lived with the gods, it was only to be reborn among humans to attain enlightenment.

Good and evil as opposites exist only within the strict framework of *saṁsāra*, the relative, the conditioned, and the illusory, but not in terms of the absolute. Good and evil, like any other constructions, are just empty concepts in absolute reality. The concept, the mental construction (*saṅkhārā*), is what singularizes the multiple, creating the idea of permanence and duration where one can find only the dependent co-arising—the impermanence of phenomena arising from multiple causes. This constitutes the third level of *dukkha*. That good and evil are devoid of reality is what makes us suffer, as this raises fundamental questions about the meaning of life and, therefore, of morality, such as why to do good rather than evil. Such questions arise only if we remain within *saṁsāra*, that is if we are not yet free from the supreme illusion. But whoever wants to reach the final extinction of suffering and to end the illusion forever must abandon not only good and evil, but also any conception of the dualism of the self, for "dualistic self-centeredness sets behavior into an inexorably centripetal course" (Hindery 2004).

What is primary, therefore, is not the destruction of good and evil but the annihilation of the construction process and of the illusion, so that the dualism of good and evil, like any other dualism, no longer appears. In fact, good and evil are empty in reality only for the Buddhas and all those like the great bodhisattvas[10] and arhats[11] who have put an end to the illusion and have attained enlightenment. Until then, for all others, doing good and not doing evil remains an ethical imperative. Even a neophyte who adheres to this principle is necessarily obliged to do good and avoid evil, not only to his human neighbors but to aliens as well.

As in the case of the dependent co-arising, the binary *kusala/akusala* may also be understood at both the relative and the absolute level. At the relative level, *kusala* is the absence of evil both for oneself and for others, thus corresponding to the first two levels of *dukkha* (attraction and repulsion) vis-à-vis oneself and others, which includes desire for existence and nonexistence. At the absolute level, *kusala* corresponds to the third level of *dukkha*, namely the conditioned status and thus ignorance itself. Good is here understood neither as something done to oneself in order to get the fruits of *karma* nor as something done to others in the context of social relations, but as that which shatters the basis of illusion. It should be understood from the perspective of *nirvāṇa*—to obtain enlightenment and perceive reality as it is beyond dualism. This understanding of the good is therefore free from all consequences of *karma* and from the concept of the self. "And what is right view? Right view, I tell you, is of two sorts: There is right view with effluents [*āsava*], siding with merit, resulting in the acquisitions [of becoming]; and there is noble right view, without effluents, transcendent, a factor of the path" (*Mahācattārīsaka-sutta*, translation by Thanissaro Bhikkhu).

Purification of Mind

The action of someone who acts from the perspective of *nirvāṇa* will be devoid of any reference to good and evil, beyond all dualism. "It is in the world of duality, where discriminating consciousness of everyday life holds sway, that we must prepare the approach to Prajna-consciousness, that super-awareness of Non-duality which is the heart of Enlightenment, and therefore the goal of all man's efforts in the field of mind" (Humphreys 1977). Such an action is devoid of the sense of the self and of self-interest. Strictly speaking it will be a "disinterested action" because *karma* is founded on the motivation of an action.

> The refraining from all that is harmful,
> The undertaking of what is skillful,
> The cleansing of one's mind—
> This is the teaching of the awakened.
>
> (*Dhammapāda*, verse 183)

This purification of the mind, however, should not be understood as an active exercise, aimed at destroying the concept of evil or even of good, but as the direct experience of reality as it is. It is through the lens of the absolute Reality that the mind gradually becomes aware of the illusion at work. Through the observation of phenomena in the appearance and disappearance of good and evil, the mind becomes accustomed to seeing how these conceptions are superimposed on Reality. Thus the purification of mind calls into question the objectification of perception—objectification understood as the contamination of mind, which leads to the formation of concepts. Such concepts gradually alienate the subject of knowledge from what is conceived as the object of knowledge. This process begins with the "creation of the self" and ends with a complete separation from the rest, though with an illusion that the rest are the "possession" of the self, as is evoked in the verse above: a gradual movement from "I am the Earth" to "I am of the Earth" and finally to "the Earth is mine."[12]

When mind is purified, consciousness is liberated from the tendency of creating concepts through a simple observation of what is as it is: the dependent co-arising, action without actor. In this way the eightfold noble path is not a path of actions leading to results. Similarly, *nirvāṇa* is not a construct, but merely the absence of any action. This means that there is no action of an actor as usually conceived in dual notions; there are only actions within the law of dependent co-arising. However, it would be incorrect to think that a Buddha, a bodhisattva, or an arhat no longer acts. The lives of those thus "liberated" confirm that they teach, in their own way, at least by example and through other multiple skillful means. Here we must speak of a "typology of the action" whose measure, according to Buddhist teaching, is *karma* and egoistic intentions. At the bottom of the ladder, in submission to the illusion of *saṁsāra*, an actor acts intentionally in order to obtain a good (*Puñña*) either for himself or for others. At the second level—still in *saṁsāra* but with the illusion slowly lifting—the envisaged good is *kusala* leading to *nirvāṇa*. At the top of the ladder, finally, when illusion is destroyed and *nirvāṇa* obtained, action (not *karma*) will be totally cleansed of ego and become "disinterested." Thus there will be neither good nor bad to consider. Without any interest, neither for oneself nor for others, this disinterested action will be supremely effective in view of *nirvāṇa* for others. Thus, those who have reached the "other side" continue to act here, but theirs are pure actions.

> There is no fear for the wide awake—
> The one who has let go of gain (good = *Puñña*) and loss (evil = *pāpa*)
> Whose mind is not moistened by passion
> Whose thoughts are unassailed.
>
> (*Dhammapāda*, verse 39)

Having nothing to fear in relation to the self, the enlightened one shows the path leading to the final liberation from suffering. This gift (*Dāna*) is, therefore, the first of all virtues, the first "perfection" (*pāramī*), which Buddhas or bodhisattvas exercise, and the gift of *dharma* surpasses all other gifts. Henceforth "beyond good and evil" (*Puññapāpapahīnassa*), the Teacher becomes teaching and his every action is teaching. The Buddha is no longer a human in the ordinary sense; he is the *Tathāgata* and *Dharmakāya*, the incarnated *dharma*.

Devoid of any reference to her own self (builder now vanished), the Enlightened One is liberated from the good/evil binary. Along with this, the notion of efficiency (the supreme good of *kusala*) also disappears. This is why teaching itself must be eventually abandoned, as a mere "raft," as must the focus on the person of Buddha, although he is the teacher par excellence, the "teacher of gods and humans." At the end of the journey, the enlightened disciple may go back, but she will see neither path nor pilgrims: the builder and her actions, concepts and methods, all disappeared along with the illusion.

From what we have discussed so far, we can draw the following conclusions for astroethics. First of all the idea of dependent co-arising is not restricted to humans alone; it extends to all beings of all worlds. That we are dependent on one another is already an affirmation of the existence of the stranger, the alien, the unknown, and the unforeseeable—to borrow the expression from Derrida—the totally other (*tout autre*). Insofar as we are dependent on them, we need to show hospitality toward them.

Second, although Buddhism rises above the traditional good/evil binary, it holds that as long as we remain in illusion in the cycle of births and deaths, doing good and avoiding evil becomes a moral imperative for those who are not initiated into the deep philosophy of life. The nobility of Buddhist ethics is at the last level of *saṁsāra*, when we become enlightened and enter into the state of liberation, for we realize there is no such thing as self. The very notion of self is the key factor of identity in opposition to difference; the other is different from me and is potentially hostile. Hence, there is a drive to do away with the other, especially if the other is significantly different from me, as in the case of beings on other planets. By pulling down the very notion of the self, Buddhism offers us a unique possibility of embracing the totally other in a far more liberated way.

Hinduism

Hinduism is vast, heterogeneous, and complex. I do not intend to give an outline of the many traditions that coexist within Hinduism, although each tradition is sometimes very different from the others. Despite these differences, even the oppositions, there is a certain unity and harmony that the Hindus call *Sanatāna-dharma* (eternal or universal harmony). Hinduism is neither a religion nor polytheism. It is a family of spiritual paths with several traditions. It is not polytheistic, for every Indian, without exception, holds there is only one Supreme Being. However, there is no one privileged revelation, as claimed by the religions of the Book. The *Ṛg-veda* (1.164.46) says: "Truth is one; Sages call it by many names." Later, Śaiva Siddhānta describes this truth as Love: "Love is God and God is One."

For the purposes of astroethics, we will study one single text, a mythical poem called *Puruṣa-sūkta* in the *Ṛg-veda* (10.90). The choice of this poem is made for several reasons: First, a text from the *Ṛg-veda* is generally acceptable to most Hindus, even if they belong to different traditions. Secondly,

Vedic texts are polysemous—rich in meanings—thus giving rise to various interpretations based on the claims of different philosophical and religious traditions within Hinduism. Thirdly, this particular text, rich with semantic wealth, has left an indelible mark in the Hindu mind, but it has also been subjected to serious criticism. Fourthly, it has exerted great influence on the successive developments of Hindu thought (Biardeau 1989). Finally, it still plays an important role in the celebrations of post-Vedic rituals (Gonda 1980). I will attempt to show how this poem offers us a possibility of constructing an astroethics, because it underscores the interrelatedness of everything in the Universe—humans and nonhumans as well.

Cosmic Correlation

The first idea that the myth suggests is a correlation not only among humans themselves, but also among humans and other forms of life. This way of perceiving reality continues to mark the Hindu ethos: "A Hindu believes that not only his brother or neighbor, but also animals, constitute humanity. He recognizes that all life is a unity" (Brown 1966). Humans and other beings belong to the same cosmic family, for all beings of the Universe have a common source from which they come into existence. In the attempt to trace a common origin—one that explains the relation between humans and other beings in the Universe—the Vedic sages have suggested a mysterious *Puruṣa* as the source of all, as a victim of cosmic sacrifice from which all life is born. Even if the Upaniṣads do not ascribe importance to the idea of sacrifice, they retain the central idea that *Puruṣa* is the author of everything and that there is a correlation between humans and other forms of life in the Universe.

The relation between humans and others is not only static, that is, in form and appearance, but also dynamic. Humans have the rhythm of the heartbeat; the Earth has the rhythm of day and night. Women have their menstrual cycles; nature has the cycle of the seasons. The rhythm of nature (*ṛta*) is essential for the well-being of humankind, to maintain order and justice (*dharma*) in society. It is important to maintain this cosmic rhythm through *yajña*, the ritual sacrifice. If the sacrificial act of *Puruṣa* is the mystery of gift, then sacrifice is the law deeply embedded in every creature; *yajña* is *prathama-dharma*—the fundamental law. The word *dharma* comes from the root meaning *dhṛ* (to hold), thus suggesting that it is the sacrifice that holds the Universe together. In the epic *Mahābhārata* (12.251.25), Bhīṣma told Yudhiṣhṭhira that *dharma* was created precisely to maintain harmony (*loka-saṅgraha*) in the Universe. The first sacrifice is the archetype, which leads us to recognize the dynamic communion between humans and nature and to strengthen this communion. The more we discover our relationship to the Universe, the more we are motivated to make our life a sacrifice for the harmony of the Universe. The human person, in order to be fully human, needs nature. Vallalār, a great Tamil poet in the Śaivite tradition, expresses this idea thus: "I wither every time I see a withering shrub." Some contemporary Indian thinkers tell us that not only are humans dependent on other forms of life for their meaningful existence, but the reverse is also true. Jeyashanthi, a Tamil novelist, expresses this interdependence and mutual reciprocity between humans and other forms of life in her novel on ecofeminism titled *Kiliyammā enra Kumanā* (Jeyashanthi 2005).

We humans are dependent on other beings in the Universe to live our humanity in its fullness. This implies that we need to be at home with and relate to every being in the Universe as brothers and sisters (*bandhutva*). To be able to transcend the human community and to adopt an ethical relationship of responsibility for other beings of the Universe, we need to cultivate the art of listening to nature. By narrating the story of the young man Satyakāma, the *Chāndogya-upaniṣad* (4.4–9) states that nature is our first teacher. Satyakāma in this story is instructed by a bull, then by fire, by a swan, and finally by a sunbird, before he is instructed by his human master (*guru*) Gautama on the ultimate reality. Gautama found Satyakāma to have already gained knowledge of the ultimate reality (*brahma-vid*). For Tagore, the more we are close to nature, the more we discover our own artistic potential: "The support of nature is more essential than ever as we grow and learn, before we plunge into a life conscious of Directors. Trees and rivers, and blue skies and beautiful views are just as necessary as benches and blackboards, books and examinations" (Tagore 1961).

A major cause of our contemporary crisis is the loss of the sense of interrelatedness and interdependency, resulting in ruthless violence of humans toward others. We humans have so alienated ourselves from the rest of the Universe that we no longer see ourselves as part of the Universe. We have created an anthropocentric world, as if everything else exists only to meet human ends. To be able to relate ethically to other beings of the Universe, we need to acknowledge the sacredness in others, sacred because any alienation of humans from them results only in the annihilation of humans, a danger to which we are exposed due to our indifference to nature, nonhuman life, and sometimes even due to a certain human arrogance of presumed mastery over every other being. We miss out on the dimension of mystery; we fail to wonder at life and the interconnectedness of all living beings, which stimulates important reflections on astroethics.

Sacrifice as Self-Giving

Puruṣa-sūkta says that *Puruṣa* is "the Lord of immortality." Therefore, he is not only the foundation of our existence, but also our liberation: the fulfillment of the deepest desire of the human person. Hinduism speaks of the three paths of liberation: *karma-yoga* (path of action), *jñāna-yoga* (path of wisdom), and *bhakti-yoga* (path of affectionate devotion). They are not mutually exclusive. *Puruṣa-sūkta* has traces of all these three paths. And why do we call these paths "yoga"? The word "yoga" comes from the verbal root *yuj*, which means to unite. *Puruṣa-sūkta* tells us that creation was possible only when *Puruṣa* was fragmented. We have, therefore, an experience of fragmented life, and yoga is an experience of gathering. In this gathering, we arrive at a deep awareness of what we really are—persons living in communion. The true *yogin* is one who sees everything in her and herself in everything (*ātma-aupamya*, *Bhagavad-gītā*, 6.32), without adhering to any form of separation (*sama-darśana*, 6.29), for all of us are fragments of the same *Puruṣa*. A true *yogin* lives in perfect communion with all beings; she resides in them (*sarva-bhūta-ātma-bhūta-ātma*, 5.7). She is the friend of all (*sarva-bhūta-suhṛt*, 5.29) and is committed to the welfare of all (*sarva-bhūta-hita-rata*, 5.25). A true *yogin* thus becomes the heart of all that exists (*sarva-bhūta-hṛdaya*) (Anand 2004).

Sacrifice (*yajña*) still remains important in the post-Vedic period, but the meaning of sacrifice becomes broader. It is no longer limited to ritual but has come to mean "self-sacrifice" for the welfare of all life in the Universe. The Vedic tradition has already insisted on the five great sacrifices (*pañca-mahạ-yajña*) in everyday life: (1) feeding all life on Earth, (2) showing hospitality to all, (3) remembering our ancestors, (4) thanking the divine forces, and (5) glorifying Brahman, the Supreme Being, through the study of sacred writings that guide us on our spiritual path (*Śatapatha-brāhmaṇa* 11.5.6.1–3). In the Upaniṣads, this understanding of sacrifice is further developed by interpreting the events of everyday life in sacrificial terms (*Chāndogya-upaniṣad* 3.16; 5.4–8). In the post-Vedic literature, the central idea of sacrifice, namely, *advaita*, nonseparated mode of being, is retained. The deeper reality within the whole of creation is the Supreme One (Brahman). The immanent principle of our existence is the divine: *ātman* is

Brahman. The divine creative act is not like that of a carpenter, who remains outside of her product. The carpenter may die, but her product continues to exist because the causal act of the carpenter determines only the structure of her product, not its existence. Brahman is the existence of everything (*satyasa satyam*, *Bṛhadarāṇyaka-upaniṣad* 2.1.20) and our deepest inner self. The presence of Brahman in us is more vital than our breath, and therefore Brahman is the breath of our breath (*Chāndogya-upaniṣad* 4.1.14). Did not St. Augustine say, "God is closer to me than I am to myself"?

The Bhagavad-gītā reaffirms that sacrifice is the source of everything, emphasizing the intrinsic link between human persons and the Universe. Anyone who does not make sacrifice, but benefits from nature, is a thief (3.10–16). The meaning of sacrifice is wider than that of ritual sacrifice. Kṛṣṇa explains that sacrifice (*yajña*) can have different meanings (4.24–32), *jñāna yajña*—wisdom as the highest sacrifice. Thanks to wisdom, we work in a spirit of service without seeking any personal gain. Wisdom purifies our work (3.19) and therefore our work becomes a sacrifice (4.22). There is a significant transition from *karma-yoga* to *jñāna-yoga* in the post-Vedic literature. Although *karma-yoga* is not entirely rejected, the meaning of *karma* is not the same: *karma* is no longer ritual sacrifice but sacrifice in the service of others.

Bhakti-yoga (affectionate devotion) has gradually gained popularity, through a mystical consciousness of the nondual (*advaitic*) relationship between the divine, the cosmos, and the human. *Bṛhadarāṇyaka-upaniṣad* (3.4.21) makes a comparison between the union of a devotee with Brahman and the union of a man with his beloved wife. The major schools of both traditions, Vedānta and Siddhānta, build their theological foundation on the concept of *advaita* to explain the erotic love between the human person and the divine, and some schools of Vedānta see in the love between Kṛṣṇa and Rādhā a model of love par excellence. This union of love is like the desire of the Shulamite woman, in the Song of Solomon, to meet her lover. This same desire is found in Andal's love of God, Kṛṣṇa, in the Vaiṣṇavite tradition (Clooney 1988). We can find parallels in the Christian tradition in the works of mystics like John of the Cross and Teresa of Avila. This mystical experience, irrespective of religious affiliations, helps us see the divine in everything.

Significance of Advaita

This genesis story highlights the reversible roles between *Puruṣa* and *Virāj* (the first woman born of *Puruṣa*). The primordial being is bipolar, but it is a dynamic polarity: one giving rise to another. *Puruṣa* is an androgyne; the common source is both male and female, *Ardhanārīśvara*, one of the many names of Śiva.[13] Thus, the poem points to the androgynous character of the cosmic man and emphasizes the indissolubility of the cosmic couple (Varenne 1982). The Jewish mystics, like the Indian mystics, believe that whatever happens in the world is a reflection of whatever happens in heaven, and vice versa. Heaven can be influenced by human behavior on Earth. "The world is unredeemed because God's masculine side is separated from God's feminine side, and when husband and wife come together in love, they restore God's unity and bring the Redemption closer" (Kushner 1997). The reversibility of roles in the creative moment expresses the truth that men and women are equal and need each other.

Bṛhadāraṇyaka-upaniṣad (1.4.1–3) narrates a myth similar to that of *Puruṣa-sūkta*:

> First and foremost, self alone was this: which is established as *Puruṣa* [the common, underlying principle of "humanness" in every person]. He looked, and saw nothing else but self. He first declared: "I am." Thus, he came to be called "I." And therefore even now, one who is addressed says first just "It is I"; and then speaks another name, which becomes his. He who came first, before all this, burned up all ills. Therefore, he is *Puruṣa* [plain, simple "humanness"]. One who knows thus, truly burns up that which seeks to come before this ["humanness"].
>
> He was afraid. Therefore, a person who is lonely feels afraid. He himself made this observation: "Since there is nothing else but me, of what then am I afraid?" From that alone his fear departed. For, of what should one be afraid? It's only from a second thing that fear arises.
>
> But still, he was not pleased. Therefore, a lonely person is not pleased. He desired a second. He became the size of a woman and man in close embrace. This very self he caused to fall divided into two. From that, husband and wife came into being. Therefore, this [personality] is a half-fragment of oneself. This is just what Yajñyavalkya truly said. Therefore, this space is just filled by a woman. With her, he came together. From that, human beings were born.[14]

If the primordial being were alone, he would have no reason to fear, but he would have no joy either. This is the paradox of his existence. Fortunately there is a third possibility—communion, which is possible only through a gift of the self. This approach subscribes to neither a monistic nor a dualistic vision of reality. The relationship between man and woman is a further illustration of the basic idea that even though we are many, we are still one. Our sexual differentiation is a reminder of the nondual (*advaita*) character of all reality: men and women are the same, yet very different from each other. If they were completely different, their union would be unthinkable and impossible. Thanks to the *advaitic* nature of reality, their union is not only possible but also desirable and meaningful. The joy of intimate union between man and woman is just a foretaste of the experience of deep happiness in the *advaitic* union between human persons and other beings.

Life as a Flow: Hermeneutical Appropriation of Symbols

For the Vedic people, fire (*Agñi*) is a symbol both of sacrifice and love. The first book of the *Ṛg-veda* begins and ends with a hymn of praise to Agni.[15] The *Ṛg-veda* 1.1.5 says thus: "May Agni the invoker, of wise intelligence, the true, of most brilliant fame, the god come with the gods." Why is Agni called wise? This question finds response in the light of another Vedic hymn called "the river of wisdom" (*Ṛg-veda* 3:10–12) that "enlightens every mind." Water enlightens us, for it reminds us of the nature of our existence as always moving toward the ultimate, like the river that flows into the great ocean. The fundamental consciousness of human contingency is the beginning of wisdom. The human person is a pilgrim in search of her true self, the ultimate reality, an idea echoed by Gautama Buddha for whom everything passes away (*sarvam antiyam*), the human tragedy being clinging to what passes away. In our pilgrimage toward our true self, Agni invokes the sacrifice of our "empirical self," in the Husserlian sense of the term.

This flow of life, referred to by the symbol of the river, reminds us of Gaṅgā (the Ganges) descending from heaven, as in the epic *Rāmāyaṇa*. It begins with an old legend of Bhagīratha practicing austerity (*tapas*) to make Gaṅgā descend from heaven to give life to his uncle, the son of Sagara who wanted to offer the sacrifice of the hundredth horse to gain the Kingdom of Indra (*Indraloka*), reduced to ashes by the *yogin* Kapila. Brahmā granted him this favor, on the condition that Śiva would agree to receive Gaṅgā on his head, for he alone could withstand her force. Śiva agreed to it, thus making Gaṅgā holy and liberation possible for all who bathe in this holy river. Gaṅgā therefore received the name Bhagirāthi (the one that gives life).

Gaṅgā is presented as the daughter of Himavāt, the mountain (*parvata*). There are two basic symbols: the mountain symbolizes stability and the river movement, which means that we should not get attached to anything: there is nothing permanent except the source (*alpha*), and we are always moving toward the goal (*omega*). Life must have two dimensions: stability and motion. Because we are always moving, we are not identical to the source, which is stable. Yet we are not entirely different from it, which recalls the principle of nonduality. There are also two other dimensions: *saṁsāra* (the flow of life) and *mokṣa* (liberation as the goal). *Saṁsāra* has meaning only in *mokṣa*, the transcendental dimension of the flow of life. It is not by coincidence that Swami Abikshitananda (Henri Le Saux) said, "My heart is now divided between the sacred river (Kavery) and the sacred mountain (Arunachala)."

The symbol of the river evokes yet another meaning. In ancient civilization, a river was a natural border, and it had to be crossed over (*Tīrtha*) to go beyond. The border is a symbol of the limits of space and time, and crossing over symbolizes eternity and thus liberation (*mokṣa*).[16] The river is also a powerful symbol of the womb giving rise to life. It is also a symbol of purity, not only physical but also ontological, dying to self-confinement to open up to others. Life is a perpetual march toward the ultimate reality, a constant going out of oneself. That is why the river is also a symbol of wisdom. The fundamental affirmation of wisdom is that everything passes, and the worst tragedy of mankind is to cling to what is fleeting. Sarasvati, the wife of Brahmā, is designated in Indian legends as the goddess of wisdom. Sarasvati, whose name comes from the root "to flow," implies that we are pilgrims toward the final goal, symbolized by the ocean into which the river flows.

In India the most sacred place is the meeting (*saṇgama*) of the three rivers Gaṅgā, Yamunā, and Sarasvati. Gaṅgā, whose root word means "the one who descends to Earth," symbolizes constant movement. Yamunā, the daughter of Yama, the lord of death, symbolizes that every being is a being unto death. Finally Sarasvati, symbol of wisdom, representing an invisible and underground river, means the possibility of overcoming death by annihilating the false self (*ahaṅkāra*) to gain liberation. Hence the importance of prayer in *Bṛhadarāṇyaka-upaniṣad* (1.3.28): "Lead me from the unreal to the Real, Lead me from darkness unto Light, Lead me from death to Immortality."

The meeting place of the rivers is sacred also because it is where the divine and humans meet. The most sacred time is dawn (*sandhyā*, "union"). The dawn comes between night and day and is associated with the horizon where the sky meets the Earth. It is a symbol of the encounter between the divine and the human, between life and death, between light and darkness, between grace and failings. It is at dawn that Hindus traditionally recite the *Gāyatrī-mantra* (*Ṛg-vēdā* 3.62.10). This prayer makes the encounter between God and man possible. When this encounter takes place, there is a profound experience of the dawn, the spiritual awakening. The Hindu man begins his spiritual journey with the sacrament of initiation (*upanayana*) and enters the first stage; he becomes *Brahmacārin* by devoting his life to the study of sacred writings. Then he gets married, builds his family, and becomes the head of the family (*gṛhastha*). In the third stage of his life, he lives in the forest; he becomes a forest dweller, *vanavāsa*. The forest is a symbol of nature in her purity, uncontaminated by the avarice of humans, having thus the possibility of communicating with the divine. Without this experience of the divine through nature, the human person cannot pass on to the final step, the life of a vagabond, *sannyāsa*. The vagabond life is a symbol of complete freedom (*mokṣa*): the human person is no longer restricted by any frontier. This freedom is not possible without the gift of the self, and this freedom also makes this self-giving possible, thus to get "united" to the Supreme Being, Infinite Consciousness, and Profound Bliss (*Sat-Cit-Ānanda*). This great "Self" is fullness itself. The Universe was born out of the fullness, but nothing is either removed or added to the fullness. "That is full; this is full. The full comes out of the full. Taking the full from the full, the full itself remains" (*Īśa-upaniṣad* 1).[17] The Universe is as sacred as fullness that is dormant in the minerals, awake in vegetation, makes animals move and humans meditate. The "mystical belief that each self or *ātman* is identical with the one *Ātman-Brahman* in the fullness of the divine goodness can lend support both to self-love and to hope in the goodness of others" (Hindery 2004).

Conclusion

Generally speaking, both Buddhism and Hinduism uphold the principle of nonduality (*advaita*). While Buddhism lays emphasis on dependent co-arising, the illusion of the self as a distinct identity, Hinduism underscores the common source of origin and interrelatedness of all beings for continued existence. Both of these ancient religious traditions are fertile terrain to take ethics beyond conventional anthropocentric ethical paradigms toward astroethics. The philosophical language of these religious traditions is deeply poetical and mythical, falling outside the scope of logical argumentation. But does not poetry rather than prose, and narrative rather than proposition, better communicate the deep mystery of life?

Notes

1. "The notion that all these fragments are separately existent is evidently an illusion, and this illusion cannot do other than lead to endless conflict and confusion. Indeed, the attempt to live according to the notion that the fragments are really separate is, in essence, what has led to the growing series of extremely urgent crises that is confronting us today" (Bohm 2002, 2).

2. Einstein as quoted in the *Zen Mountain Monastery Newsletter* of 1989. This quote is not found in any of Einstein's original wirings. However, his writings suggest that his thought reached beyond a personal God and that he had an appreciation for Buddhism.

3. Prasad 2007, 43–50, gives a brief background on Buddhist cosmology.

4. The Sanskrit language maintains a difference between the written form and the oral articulation of some words. One such word is "Brahman." In the written form, "Brahman" in the neuter gender refers to the Supreme Being and it is pronounced as such. But "Brahman" in the masculine gender refers to a minor god, who is the creator. Though it is written as "Brahman," it is pronounced Brahmā. To avoid confusion, in this chapter I shall spell the word as it is pronounced, whenever we refer to Brahman in the masculine gender.

5. *Paṭiccasamuppāda* is a Pāli word translated as dependent co-arising or dependent origination. By "dependent co-arising" we understand that every phenomenon is constructed and originates from multiple causes that are mutually dependent: if the tree is born of the seed, its birth is also made possible by the land, oxygen, water, and space for it to unfold. There is no single origin, not even for freedom: a phenomenon occurs only with the meeting of the multiple conditions that permit it. To say that the tree is born of the seed is an arbitrary simplification, a figment of imagination. This idea is central to Buddhist teachings, particularly to Buddhist ethics. Everything is mutually causative, whether persons or things, and morality is grounded in this interdependence. See Joanna Rogers Macy, "Dependent Co-Arising: The Distinctiveness of Buddhist Ethics," *Journal of Religious Ethics* 7, no. 1 (Spring 1979): 38–52.

6. *Kusala* and *akusala*, though commonly translated as good and evil in English, refer to well-being and ill-being, respectively. See Payutto 1993, 17–18.

7. Serving as an everyday moral manual, it sets out in short verses the basic rules for practicing Buddhism. Extremely popular, it is the most widely read book, the best known among the population of Theravada countries of Southeast Asia.

8. The first two opening verses of *Dhammapāda* clearly show the opposition between good and evil. See verses 1–2, 15–18.

9. If the word *karma* is translated etymologically as action, it comes from the Indo-European root *kṛ*, which in English corresponds to the word "create." *Karma* is therefore a creation, a construction, a voluntary and intentional act in favor of the idea of the self. Insofar as an action perpetuates the idea of self, through the binary desire/aversion, whatever we do nourishes the idea of self and causes rebirth.

10. A bodhisattva is a being destined to become a future Buddha. Though enlightened, he needs to be born for the last time as Buddha.

11. "Arhat" also means enlightened being. Radhakrishnan (1948, 428, footnote 2), among others, suggests that this term was used to denote enlightened being even before the pre-Buddhist period.

12. See "The Division of the Discourse on the Root."

13. This idea of God as male and female is also found in many ancient traditions, including the Kabbalah of Judaism.

14. A similar myth exists in Greek mythology. Aristophanes makes use of this myth in Plato's *Symposium*, to give a reason for human erotic desire.

15. The two basic symbols in the Vedas are fire and water. The same is true of Christianity: the fire of the Holy Spirit and the water of baptism. These are probably universal symbols.

16. The same image is also evoked in the book of Exodus in the passage of the Red Sea, an image that points to the paschal mystery, passage par excellence for a Christian from death to life.

17. The original Sanskrit text reads as follows: *pūrṇamadaḥ, pūrṇamidam, pūrṇātpūrṇamudacyate, pūrṇasya pūrṇamādāya, pūrṇamevāvaśiṣyate.*

References

Anand, S. 2004. *Hindu Inspiration for Christian Reflection: Towards a Hindu-Christian Theology*. Anand: Gujarat Sahitya Prakash.

Aurelius, M. 1998. *Meditations*. Translated by Arthur Spenser, L. Farquharson, and R. B. Rutherford. Oxford: Oxford University Press.

Biardeau, M. 1989. *Hinduism: The Anthropology of a Civilization*. Oxford: Oxford University Press.

Bohm, D. 2002. *Wholeness and the Implicate Order*. London: Routledge.

Brown, N. 1966. *Man in the Universe: Some Continuities in Indian Thought*. Berkeley: University of California Press.

Buddhaghosa. 2010. *Visuddhimagga* [*The Path of Purification*]. Translated by Bhikkhu Ñānamoli Colombo. Sri Lanka: Buddhist Publication Society.

Clooney, F. X. 1988. *From Hindu Wisdom for All God's Children*. New York: Maryknoll.

Dhammapāda. 2007. Translated by Glenn Wallis. New York: Random House.

"The Division of the Discourse on the Root" (Mūlapariyāyavagga). In *Majjhima-nikāya*, pt. 1, Mūlapannāsapāli, §3. Translated by Bhikkhu Ñāṇamoli and Bhikkhu Bodhi. Accessed May 24, 2012. www.palicanon.org/en/sutta-pitaka/transcribed-suttas/majjhima-nikaya/62-mn-1-mlapariyya-sutta-the-root-of-all-things.html.

du Pre, G. 1984. "The Buddhist Philosophy of Science." In *Buddhism and Science*, edited by Buddhasada P. Kirthisinghe. Delhi: Motilal Banarsidass.

Gonda, J. 1980. *Vedic Ritual: The Non-Solemn Rites*. Leiden: Brill.

"The Great Forty" (Mahācattārīsaka Sutta). In *Majjhima-nikāya*, pt. 3, Uparipannāsapāli, §117. Translated by Thanissaro Bhikkhu. Accessed May 24, 2012. www.accesstoinsight.org/tipitaka/mn/mn.117.than.html.

Hindery, R. 2004. *Comparative Ethics in Hindu and Buddhist Traditions*. Delhi, India: Motilal Banarsidass.

Humphreys, C. 1977. *The Buddhist Way of Action*. London: Allen and Unwin.
Jeyashanthi, J. I. 2005. *Kiliyammāenra Kumanā*. Chennai: Madhi Nilayam.
Kalupahana, D. J. 1995. *Ethics in Early Buddhism*. Delhi: Motilal Banarsidass.
Kushner, H. S. 1997. *How Good Do We Have To Be?* Boston: Little, Brown.
Leibniz, G. W. 1965. *Philosophical Writings*. Translated by M. De S. Morris. London: Dent. Originally published in 1670.
Payutto, P. A. (Bhikku). 1993. *Good, Evil and Beyond: Kamma in the Buddha's Teaching*. Translated by Bhikku Puriso. Bangkok, Thailand: Buddhadhamma Foundation.
Prasad, H. S. 2007. *The Centrality of Ethics in Buddhism*. Delhi: Motilal Banarsidass.
Radhakrishnan, S. 1948. *Indian Philosophy, Vol. I*. London: George Allen and Unwin, Ltd.
Spinoza, B. 2009. *Ethic: Demonstrated in Geometrical Order and Divided into Five Parts*. Translated by W. H. White. Charleston, SC: Charleston BiblioBazaar. Originally published in 1673.
Tagore, R. 1961. *Towards Universal Man*. Bombay: Asia Publications.
Thuan, Trinh Xuan. 2005. *The Secret Melody: And Man Created the Universe*. West Conshohocken, PA: Templeton Press.
Varenne, J. 1982. *Cosmogonies védiques*. Paris: Les Belles Lettres.
Vasubandhu. 1990. *Abhidarmakośabhāsyam*. Vol. 1. Translated by L. M. Pruden. Berkeley, CA: Asian Humanities Press.

CHAPTER SIX

Social and Ethical Implications of Creating Artificial Cells

Mark A. Bedau
REED COLLEGE

Mark Triant
319 SCHOLES GALLERY

Artificial Cells and Astrobiology

The astrobiology search for extraterrestrial life and the prospect of someday actually encountering extraterrestrial life are making us rethink certain ethical and social issues. Many of those issues arise in part specifically because extraterrestrial (ET) life could be so different from any of the myriad forms of life we know about today. There is a second scientific context in which extremely unfamiliar forms of life generate similar societal and ethical issues: scientists making artificial cells, or as they are sometimes called, protocells (Rasmussen et al. 2009). Artificial cells are complex chemical systems that are created in certain special laboratory conditions. Scientists are intentionally pushing the boundaries of minimal forms of chemical life, simplifying as much as possible and using whatever materials they find convenient or interesting, exploring the minimal chemical conditions of life-as-it-could-be. Creating these artificial cells raises a number of deep and controversial social and ethical issues. This chapter explores those issues.

* This selection is reprinted with permission from *The Ethics of Protocells: Moral and Social Implications of Creating Life in the Laboratory*, edited by Mark A. Bedau and Emily C. Parke, Cambridge, MA: MIT Press, 2009, and modified for this volume by Mark Bedau.

Astrobiology also puts pressure on the boundaries of minimal chemical life. Extraterrestrial life could be very unfamiliar and could exist under chemical conditions very different from those found on Earth (Pace 2001; Benner et al. 2004). The unfamiliarity of extraterrestrial life makes the outcome of any encounter with it extremely uncertain, and this uncertainty creates certain dangers and risks for society (see Benner and Woolf in the appendix of this volume for further discussion of these risks, and the chapters by Race and Sullivan on closely related issues). In this respect extraterrestrial life and artificial cells are similar. Both have highly unpredictable consequences, so the proper stance toward each requires similar precaution. This example illustrates some of the connections among the broader issues raised by artificial cells and extraterrestrial life.

In other respects, though, artificial cells and extraterrestrial life raise broader issues that differ. For example, artificial cells are intentionally, consciously created by humans, and this opens the door to ethical concerns about intentional abuse or malicious use. Because humans do not create extraterrestrial life-forms, they cannot create them maliciously, but they could still use them for malicious purposes. So, human intentions have different but somewhat similar effects on the broader implications of artificial cells and extraterrestrial life. A further difference between artificial cells and extraterrestrial life is commercial potential and intent. Artificial cell research is partly driven by pure scientific curiosity, but another dominant driver is potential practical applications. The ability to program artificial cells to do our bidding creates great economic opportunities and has great potential value for society. It also creates various risks and threats. In contrast, the practical and commercial fruits of astrobiology seem much more narrow and limited.

Because of both the similarities and the differences, a thorough investigation of the social and ethical implications of astrobiology would include a look at the implications of artificial cells. We learn something from both the analogies and the contrasts.

What Are Artificial Cells?

A striking biotechnology research program has been quietly making incremental progress for the past generation, but it will soon become public knowledge. One small sign of this is a 2002 article in the widely distributed Sunday supplement *Parade Magazine*, in which one could read the following prediction: "Tiny robots may crawl through your arteries, cutting away atherosclerotic plaque; powerful drugs will be delivered to individual cancer cells, leaving other cells undamaged; teeth will be self-cleaning. Cosmetically, you will change your hair color with an injection of nanomachines that circulate through your body, moving melanocytes in hair follicles" (Crichton 2002a). This may sound incredible, and it is certainly science fiction today, but scientists working in the field believe that within the next decade or so the basic technology underlying these predictions will exist. That technology could be called "artificial cells."

Artificial cells are microscopic self-organizing and self-replicating autonomous entities created artificially from simple organic and inorganic substances. The *Parade* article quoted above was written by Michael Crichton as an effort to promote his recent bestseller, *Prey* (Crichton 2002b). Crichton's book imagines the disastrous consequences of artificial cell commercialization gone awry (swarms of artificial cells prey on humans). Although one can question many scientific presuppositions behind Crichton's story (Dyson 2003), research to create artificial cells is proceeding apace, and it is fair to say that the potential risks and benefits to society are enormous. When pictures of Dolly, the Scottish sheep cloned from an adult udder cell, splashed across the front pages of newspapers around the world, society was caught unprepared. President Clinton immediately halted all federally funded cloning research in the United States (Brannigan 2001), and polls revealed that 90 percent of the public favored a ban on human cloning (Silver 1998). To prevent the future announcement of the first artificial cells from provoking similar knee-jerk reactions, we should start thinking through the implications today.

This chapter reviews many of the main strategies for deciding whether to create artificial cells. One set of considerations focuses on intrinsic features of artificial cells. These include the suggestions that creating artificial cells is unnatural, that it treats life as a commodity, that it promotes a mistaken reductionism, and that it is playing God. We find all these considerations unconvincing, for reasons we explain below. The alternative strategies focus on weighing the consequences of creating artificial cells. Utilitarianism and decision theory promise scientifically objective and pragmatic methods for deciding what course to chart. Although we agree that consequences have central importance, we are skeptical whether utilitarianism and decision theory can provide much practical help, because the consequences of creating artificial cells are so uncertain. The critical problem is to find some method for choosing the best course of action in the face of this uncertainty. In this setting some people advocate the doomsday principle, but we explain why we find this principle incoherent.

An increasing number of decision makers are turning for guidance in such situations to the precautionary principle, but we explain why we find this principle also unattractive. (Cleland, Stoeger, and Sullivan extensively discuss the ethical foundations and the intrinsic ethical status of life and ET life elsewhere in this volume. This chapter focuses on criteria that apply directly to risks and benefits, which complements the discussion by Benner and Woolf in the appendix.) We conclude that making decisions about artificial cells requires being courageous about accepting uncertain risks when warranted by the potential gains.

The Intrinsic Value of Life

Arguments about whether it is right or wrong to develop a new technology can take either of two forms (Reiss and Straughan 1996; Comstock 2000). Extrinsic arguments are driven by the technology's consequences. A technology's consequences often depend on how it is implemented, so extrinsic arguments do not usually produce blanket evaluations of all possible implementations of a technology. Presumably, any decision about creating a new technology should weigh the technology's consequences, perhaps along with other considerations. Evaluating extrinsic approaches to decisions about artificial cells is the subject of the two subsequent sections. In this section we focus on intrinsic arguments for or against a new technology. Such intrinsic arguments are driven by the nature of the technology itself, yielding conclusions pertinent to any implementation of it. The advances in biochemical pharmacology of the early twentieth century and more recent developments in genetic engineering and cloning have been criticized on intrinsic grounds, for example. Such criticisms include injunctions against playing God, tampering with forces beyond our control, or violating nature's sanctity; the prospect of creating artificial cells raises many of the same kinds of intrinsic concerns. Is there something intrinsically objectionable about creating artificial life-forms?

Reactions to the prospect of synthesizing new forms of life range from fascination to skepticism and even horror. Everyone should agree that the first artificial cell will herald a scientific and cultural event of great significance, one that will force us to reconsider our place in the cosmos (much as the first contact with extraterrestrial life will). But what some would hail as a technological milestone, others would decry as a new height of scientific hubris. The "Frankencell" tag attached to Venter's minimal genome project reveals the uneasiness generated by this prospect. So it is natural to ask whether taking this big step would be crossing some forbidden line. There are four kinds of intrinsic objections to the creation of artificial cells, all of which frequently arise in debates over genetic engineering and cloning. These arguments all stem from the notion that life has a certain privileged status and should in some respect remain off-limits from human intervention and manipulation.

One objection to creating artificial cells is simply that doing so would be unnatural and, hence, unethical. The force of such arguments depends on what is meant by "unnatural" and why the unnatural is wrong. At one extreme, one could view all human activity and its products as natural, because we are part of the natural world. But then creating artificial cells would be natural, and this objection would have no force. At the other extreme, one could consider all human activity and its products as unnatural, defining the natural as what is independent of human influence. But then the objection would deem all human activities to be unethical, which is absurd. So the objection will have any force only if "natural" is interpreted in such a way that we can engage in both natural and unnatural acts and the unnatural acts are intuitively wrong. But what could that sense of "natural" be? One might consider it "unnatural" to intervene in the workings of other life-forms. But then the unnatural is not in general wrong; far from it. For example, it is surely not immoral to hybridize vegetable species or to engage in animal husbandry. And the stricture against interfering in life-forms is not something that arises especially for humans, for it is not generally thought to be wrong to vaccinate one's children. So there is no evident sense of "unnatural" in which artificial cells are unnatural and the unnatural is intrinsically wrong.

Another objection is that to create artificial life-forms would lead to the commodification of life, which is immoral.[1] Underlying this objection is the notion that living things have a certain sanctity or otherwise demand our respect, and that creating them undermines this respect. The commodification of life is seen as analogous to the commodification of persons, a practice most of us would find appalling. By producing living artifacts, one might argue, we would come to regard life-forms as one among our many products, and thus valuable only insofar as they are useful to us. This argument is easy to sympathize with, but is implausible when followed to its conclusion. Life is after all one of our most abundant commodities. Produce, livestock, vaccines, and pets are all examples of life-forms that are bought and sold every day. Anyone who objects to the commodification of an artificial single-celled organism should also object to the commodification of a tomato. Furthermore, creating, buying, and selling life-forms does not prevent one from respecting those life-forms. Many family farmers, for example, are among those with the greatest reverence for life.

The commodification argument reflects a commonly held sentiment that life is special somehow, that it is wrong to treat it with no more respect than we give the rest of the material world. It can be argued that while it is not inherently wrong to commodify living things, it is still wrong to create life from nonliving matter, because doing so would foster a reductionistic attitude toward life, which undermines the sense of awe, reverence, and respect we owe it.[2] This objection does not exactly require that biological reductionism be false, but merely that it be bad for us to view life reductionistically. Of course, it seems somewhat absurd to admit the truth of some form of biological reductionism while advocating an antireductionist worldview on moral grounds. If living things are really irreducible to purely physical systems (at least in some minimal sense), then creating life from nonliving chemicals would presumably be impossible, so the argument is moot. By the same token, if living things are reducible to physical systems, it is hard to see why fostering reductionistic beliefs would be unethical. It is by no means obvious that life per se is the type of thing that demands the sense of awe and respect this objection is premised upon, but even if we grant that life deserves our reverence, there is no reason to assume that this is incompatible with biological reductionism. Many who study the workings of life in a reductionistic framework come away from the experience with a sense of wonder and an enhanced appreciation and respect for their object of study. Life is no less amazing by virtue of being an elaborate chemical process. In fact, only after we began studying life in naturalistic terms have we come to appreciate how staggeringly complex it really is.

Inevitably, the proponents and eventual creators of artificial cells will have to face up to the accusation that they are *playing God*.[3] The playing-God argument can be fleshed out in two ways: It could be the observation that creating life from scratch creates new dangers that we simply are not prepared to handle, or it could be the claim that, for whatever reason, creating life from scratch crosses a line that humans simply should never cross. The former construal concerns the potential bad consequences of artificial cells, so it will be discussed in a subsequent section.

If creating artificial cells is crossing some line, we must ask exactly where the line is and why we should not cross it. What exactly would be so horrible about creating new forms of life from scratch? If we set aside the *consequences* of doing this, we are left with little to explain why crossing that line should be forbidden.

The term "playing God" was popularized in the early twentieth century by Christian Scientists in reaction to the variety of advances in medical science taking place at the time. With the help of new surgical techniques, vaccines, antibiotics, and other pharmaceuticals, the human life span began to extend, and many fatal or otherwise untreatable ailments could now be easily and reliably cured. Christian Scientists opted out of medical treatment on the grounds that it is wrong to "play God"—healing the ill was God's business, not ours. Yet if a person living today were to deny her ailing child medical attention on the grounds that doing so is playing God, we would be rightly appalled. So, if saving a life through modern medicine is playing God, then playing God is morally required.

Questions surrounding the playing-God argument are related to the more general question of how religious authority should influence scientific policy decisions. Though religious doctrine will surely be invoked in future discussions of artificial cell science, and though religious doctrine might help guide policy makers, in modern nonsectarian democratic states, religious doctrine itself has a decreasing role in shaping public policy.

All of the intrinsic objections against the creation of artificial cells canvassed in this section turn out to be vague, simplistic, or ill-conceived. So, in the next section we examine extrinsic objections that turn on the *consequences* of creating artificial cells.

Evaluating the Consequences

Policy makers will soon have to face questions about whether and under what conditions to create artificial cells. Should artificial cells be developed for commercial, industrial, or military applications? How strictly should they be regulated? The reasons for pursuing artificial cell research and development are matched by concerns about their safety. Especially this early in the decision-making process, there is no obvious way of knowing whether the speculative benefits are worth pursuing, given the speculative risks. Ruling out any intrinsic ethical qualms against creating artificial cells, the choices we make will be for the most part a matter of how we weigh the risks and benefits of artificial cells, and what strategies and principles we employ when making decisions in the face of uncertain consequences (see also Benner and Woolf's contribution in the appendix, and the related discussions in the chapters by Cleland, Race, Stoeger, and Sullivan).

The utilitarian calculus is one obvious tool we might employ in assessing a course of action in light of its consequences: Possible actions are measured according to the overall "utility" they would produce (such as the net aggregate happiness or well-being that would result), where the course of action we *should* pursue is the one that would produce the greatest utility. There are, of course, problems with any utilitarian calculus, one being the question of how to quantify a given outcome. Is it worse for a person to die or for a corporation to be forced to lay off 10,000 employees? At what point do benefits to public health outweigh ecological risks? The answers to these questions only become less clear when realistically complex situations are taken into account, but this is not a problem peculiar to utilitarianism. Normal acts of deliberation often require us to evaluate and compare the possible outcomes of our actions, no matter how different the objects of comparison. Though one may balk at the notion of assigning a monetary equivalent to the value of a human life, practical social institutions face this task every day. Whether or not the value of money is comparable to the value of a human life in any objective sense, certain acts of deliberation require that their values be compared insofar as their values enter into deciding between mutually exclusive courses of action.

This problem of comparing apples to oranges is not the only obstacle faced by the utilitarian approach. In its simplest forms, utilitarianism is insensitive to the distribution of risks and benefits resulting from a course of action. One outcome is considered better than another only if it possesses the greater *aggregate* utility. Imagine, for instance, that an experimental batch of artificial cells could be developed in the field in rural Asia and eventually save tens of thousands of people throughout Asia, but also imagine that during development the artificial cells would accidentally infect and kill 10 people. If our choice were limited to whether or not to release the experimental artificial cells, a simple utilitarian calculus would require developing the new drug. Now imagine that a commercial manufacturer of artificial cells makes a fortune off one of its products, but in doing so causes significant damage to the environment. Depending on the details of the situation and the relative value of things like corporate wealth and environmental health, the utilitarian calculus may require the commercialization of artificial cells. If this consequence seems unjust, the utilitarian rule could be adjusted to take proper account of the distribution of harms and benefits, along with their quantity and magnitude.

Beyond the question of distribution, we might want other factors to influence how harms and benefits are weighted. For instance, many consider the ongoing process of scientific discovery to be crucial for society's continued benefit and enrichment, so that when in doubt about the consequences of a given research program, we should err on the side of free inquiry.

Another plausible way one might bias one's assessment of risks and benefits would be to err on the side of the avoidance of harm. We normally feel obligated not to inflict harm on others, even if we think that human welfare overall would increase as a result. A surgeon should never deliberately kill one of her patients, even if doing so would mean supplying lifesaving organ transplants to five other dying patients. In fact, the Hippocratic oath for doctors says that, first, one should do no harm. So if, all else being equal, we should avoid doing harm, it is sensible to bias one's assessment of harms and benefits in favor of playing it safe.

Though it may be true that we have a special obligation against doing harm, a harm-weighted principle of risk assessment is unhelpful in deciding how to proceed with a program as long-term and uncertain as artificial cell science. As Stephen Stich observes, "The distinction between doing good and doing harm presupposes a notion of the normal or expected course of events" (Stich 1978). In the surgeon's case, killing a patient to harvest his organs is clearly an instance of a harm inflicted, whereas allowing the other five patients to die due to a scarcity of spare organs is not. But when deciding the fate of artificial cell science, neither pursuing nor banning this research program could be described as inflicting a harm, because there is no way of knowing (or even of making an educated guess about) the kind of scenario to compare the outcome against. In the end, artificial cells may prove to be too difficult for humanity to control, and their creation may result in disaster. But it could also be that by banning artificial cell technology we rob ourselves of the capability to withstand some other kind of catastrophe (such as crop failure due to global warming or an antibiotic-resistant plague). The ethical dilemma posed to us by artificial cells is not one that involves deciding between alternative *outcomes*; it requires that we choose between alternative *standards of conduct* where the outcome of any particular course of action is at best a matter of conjecture. So in the present case, the imperative against doing harm is inapplicable.

The ethical problem posed by artificial cells is fundamentally speculative, and cannot be solved by simply weighing good against bad and picking the choice that comes out on top. We can at best weigh *hypothetical* risks against *hypothetical* benefits and decide on the most prudent means of navigating this uncertainty.

This still leaves room for something like a utilitarian calculus, however. Decision theory formulates principles for choosing among alternative courses of action with uncertain consequences, by appropriately weighing risks and benefits. Given a particular decision to make, one constructs a "decision tree" with a branch for each candidate decision, and then sub-branches for the possible outcomes of each decision (the set of outcomes must be mutually exclusive and exhaustive). A utility value is then assigned to each possible outcome (sub-branch). If the probabilities of each possible outcome are known or can be guessed, they are assigned to each sub-branch and the situation is called a decision "under risk." Decisions under risk are typically analyzed by calculating the expected value of each candidate decision (averaging the products of the probabilities of each possible outcome and their utilities), and then recommending the choice with the highest expected value.

If some or all of the probabilities are unknown, then the situation is called a decision "under uncertainty." Various strategies for analyzing decisions under uncertainty have been proposed. For example, the risk-averse strategy called "minimax" recommends choosing whatever leads to the best of the alternative worst-case scenarios. The proper analysis of decisions under uncertainty is not without controversy, but plausible strategies can often be found for specific kinds of decision contexts (Resnick 1987). In every case, decisions under both risk and uncertainty rely on comparing the utilities of possible outcomes of the candidate choices.

Decisions under risk or uncertainty should be contrasted with a third kind of decision—what we will term *decisions in the dark*—that are typically ignored by decision theory. Decisions in the dark arise when those facing a decision are substantially ignorant about the consequences of their candidate choices. This ignorance has two forms. One concerns the set of possible outcomes of the candidate choices; this prevents us from identifying the sub-branches of the decision tree. The other is ignorance about the utility of the possible outcomes; this prevents us from comparing the utilities of different sub-branches. In either case, decision theory gets no traction and has little, if any, advice to offer on how to make the decision.

New and revolutionary technologies like genetic engineering and nanotechnology typically present us with decisions in the dark. The unprecedented nature of these innovations makes their future implications extremely difficult to forecast. The social and economic promise is so huge that many public and private entities have bet vast stakes on the bio-nano future, but at the same time the imagined risks are generating growing alarm (Joy 2000). Even though we are substantially ignorant about their likely consequences, we face choices today about whether and how to support, develop, and regulate them. We have to make these decisions in the dark.

The same holds for decisions about artificial cells. We can and should speculate about the possible benefits and risks of artificial cell technology, but the fact remains that we now have substantial ignorance about their consequences. Statistical analyses of probabilities are consequently of little use. So, decisions about artificial cells are typically decisions in the dark. Thus, utilitarianism and decision theory and other algorithmic decision support methods have little, if any, practical value. Any decision-theoretic calculus we attempt will be limited by our current guesses about the shape of the space of consequences, and in all likelihood our picture of this shape will substantially change as we learn more.

This does not mean that we cannot make wise decisions; rather, it means that deciding will require the exercise of good judgment. Most of us have more or less well-developed abilities to identify relevant factors and take them into account, to discount factors likely to appeal to our own self-interest, and the like. These methods are fallible and inconclusive, but when we are deciding in the dark, they generally are all we have available. It might be nice if we could foist the responsibility for making wise choices onto some decision algorithm like utilitarianism or decision theory, but that is a vain hope today.

Deciding in the Dark

Even though the consequences of creating artificial cells will remain uncertain for some time, scientific leaders and policy makers still will have to face decisions about whether to allow them to be created, and under what circumstances. And as the science and technology behind artificial cells progresses, the range of these decisions will grow.[4] The decisions will include whether to permit various lines of research in the laboratory, where to allow various kinds of field trials, whether to permit development of various commercial applications, how to assign liability for harms of these commercial products, whether to restrict access to research results that could be used for nefarious purposes, and so on. The uncertainty about the possible outcomes of these decisions does not remove the responsibility for taking some course of action, and the stakes involved could be very high. How should one meet this responsibility to make decisions about artificial cells in the dark?

When contemplating a course of action that could lead to a catastrophe, many people conclude that it is not worth the risk and instinctively pull back. This form of reasoning illustrates what we call the *doomsday principle*, which says: *Do not pursue a course of action if it might lead to a catastrophe.*[5] Something like this principle is employed by many people in the nanotechnology community. For example, Merkle (1992, 292) thinks that nanomachines that replicate themselves and evolve in a natural setting pose potential risks that are so great that the nanomachines should not only not be constructed, they should not even be designed. He concludes that to achieve compliance with this goal will involve enculturating people to the idea that "there are certain things that you just *do not do*" (emphasis in original). This illustrates doomsday reasoning that absolutely forbids crossing a certain line because it might lead to a disaster.

A little reflection shows that the doomsday principle is implausible as a general rule, because it would generate all sorts of implausible prohibitions. Almost any new technology could under some circumstances lead to a catastrophe, but we presumably do not want to ban development of technology in general. To dramatize the point, notice there is always at least some risk that getting out of bed on any given morning *could* lead to a catastrophe. Maybe you will be hit by a truck and thereby be prevented from discovering a critical breakthrough for a cure for cancer; maybe a switch triggering the world's nuclear arsenal has been surreptitiously left beside your bed; and so on. These consequences are completely fanciful, of course, but they are still possible. The same kind of consideration shows that virtually every action could lead to a catastrophe and therefore would be prohibited by the doomsday principle. *Not* creating artificial cells could lead to a disaster because there could be some catastrophic consequence that society could avert only by developing artificial cells, if, for example, artificial cells could be used to cure heart disease. Because the doomsday principle prohibits your action no matter what you do, the principle is incoherent.

The likelihood of triggering a nuclear reaction by getting out of bed is negligible, of course, whereas the likelihood of self-replicating nanomachines wreaking havoc might be higher. With this in mind, one might try to resuscitate the doomsday principle by modifying it so that it is triggered only when the likelihood of catastrophe is non-negligible. But there are two problems with implementing such a principle. First, the threshold of negligible likelihood is vague and could be applied only after being converted into some precise threshold (such as probability 0.001). But any such precise threshold would be arbitrary and hard to justify. Second, it will often be impossible to ascertain the probability of an action causing a catastrophe with anything like the requisite precision. For example, we have no way at present of even estimating if the likelihood of self-replicating nanomachines causing a catastrophe is above or below 0.001. Estimates of risks are typically based on three kinds of evidence: toxicological studies of harms to laboratory animals, epidemiological studies of correlations in existing populations and environments, and statistical analyses of morbidity and mortality data (Ropeik and Gray 2002). We lack even a shred of any of these kinds of evidence concerning self-replicating nanomachines, because they do not exist yet.

When someone proposes to engage in a new kind of activity or to develop a new technology today, typically this is permitted unless and until it has been shown that some serious harm would result.[6] Think of the use of cell phones, the genetic modification of foods, the feeding of offal to cattle. In other words, a new activity is innocent until proven guilty. Anyone engaged in the new activity need not first prove that it is safe; rather, whoever questions its safety bears the burden of proof for showing that it really is unsafe. Furthermore, this burden of proof can be met only with scientifically credible evidence that establishes a causal connection between the new activity and the supposed harm. It is insufficient if someone suspects or worries that there might be such a connection, or even if there is scientific evidence that there *could* be such a connection. The causal connection must be credibly established before the new activity can be curtailed.

This approach to societal decision making has, in the eyes of many, led to serious problems. New activities have sometimes caused great damage to human health or the environment before sufficient evidence of the cause of these damages had accumulated. One notorious case is thalidomide, which was introduced in the 1950s as a sleeping pill and to combat morning sickness, and was withdrawn from the market in the early 1960s when it was discovered to cause severe birth defects (Stephens and Brynner 2001). This perception has fueled a growing attraction to the idea of shifting the burden of proof and exercising more caution before allowing new and untested activities—treating them as guilty until proven innocent. This approach to decision making is now widely known as the *precautionary principle*: *Do not pursue a course of action that might cause significant harm, even if it is uncertain whether the risk is genuine.* Different formulations of the precautionary principle can have significantly different pros and cons (Parke and Bedau 2009). Here we will concentrate just on the following elaboration (Geiser 1999), which is representative of many of the best-known statements of the principle:[7]

> The precautionary principle asserts that parties should take measures to protect public health and the environment, even in the absence of clear, scientific evidence of harm. It provides for two conditions. First, in the face of scientific uncertainties, parties should refrain from actions that might harm the environment, and, second, that the burden or proof for assuring the safety of an action falls on those who propose it.

The precautionary principle is playing an increasing role in decision making around the world. For example, the contract creating the European Union appeals to the principle, and it governs international legal arrangements such as the United Nations Biosafety Protocol.[8] The precautionary principle is also causing a growing controversy.[9]

We are skeptical of the precautionary principle. It is only common sense to exercise due caution when developing new technologies, but other considerations are also relevant. We find the precautionary principle to be too insensitive to the complexities presented by deciding in the dark. One can think of the precautionary principle as a principle of inaction, recommending that, when in doubt, leave well enough alone. It is sensible to leave well enough alone only if things are well at present and if it is fairly clear that they will remain well by preserving the status quo. But these presumptions are often false, and this causes two problems for the precautionary principle.

Leaving well enough alone might make sense if the world were unchanging, but this is manifestly false. The world's population is continuing to grow, especially in poor and relatively underdeveloped countries, and this is creating problems that will not be solved simply by ignoring them. In the developed world, average longevity has been steadily increasing; over the last hundred years in the United States, for example, life expectancy has increased more than 50 percent (Wilson and Crouch 2001). Today heart disease and cancer are far and away the two leading causes of death in the United States (Ropeik and Gray 2002). Pollution of drinking water is another growing problem. There are estimated to be 100,000 leaking underground fuel storage tanks in the United States, and a fifth of these are known to have contaminated groundwater; a third of the wells in California's San Joaquin Valley have been shown to contain 10 times the allowable level of the pesticide DBCP (1,2, Dibromo-3-chloropropane was banned from use in 1979) (Ropeik and Gray 2002). These few examples illustrate that there is a continual evolution in the key issues that society must confront. The precautionary principle does not require us to stand immobile in the face of such problems.

But it prevents us from using a method that has not been shown to be safe. So the precautionary principle ties our hands when we face new challenges.

This leads to a second, deeper problem with the precautionary principle. New procedures and technologies often offer significant benefits to society, many of which are new and unique. Cell phones free long-distance communication from the tether of land lines, and genetic engineering opens the door to biological opportunities that would never occur without human intervention. Whether or not the benefits of these technologies outweigh the risks they pose, they do have benefits. But the precautionary principle ignores such benefits. To forego these benefits causes a harm—what one might call a "harm of inaction." These harms of inaction are opportunity costs, created by the lost opportunities to bring about certain new kinds of benefits. Whether or not these opportunity costs outweigh other considerations, the precautionary principle prevents them from being considered at all. That is a mistake.

These considerations surfaced at the birth of genetic engineering. The biologists who were developing recombinant DNA methods suspected that their new technology might pose various new kinds of risks to society, so the U.S. National Academy of Science impaneled some experts to examine the matter. They quickly published their findings in *Science* in what has come to be known as the "Moratorium" letter, and recommended suspending all recombinant DNA studies "until the potential hazards . . . have been better evaluated" (Berg et al. 1974). This is an early example of precautionary reasoning. Recombinant DNA studies were suspended even though no specific risks had been scientifically documented. Instead it was thought that there *might* be such risks and that this was enough to justify a moratorium even on recombinant DNA *research*. (See also, once again, Benner and Woolf in the Appendix of this volume. With regard to extraterrestrial life, Race and Sullivan have provided related discussions in their chapters).

The Moratorium letter provoked the National Academy of Science to organize a conference at Asilomar the following year, with the aim of determining under what conditions various kinds of recombinant DNA research could be safely conducted. James Watson, who signed the Moratorium letter and participated in the Asilomar conference, reports having had serious misgivings about the excessive precaution being advocated, and writes that he "now felt that it was more irresponsible to defer research on the basis of unknown and unquantifiable dangers. There were desperately sick people out there, people with cancer or cystic fibrosis—what gave us the right to deny them perhaps their only hope?" (Watson 2003). Watson is here criticizing excessive precautionary actions because they caused harms of inaction.

Some harms of inaction are real and broad in scope, as the threat of rampant antibiotic-resistance can illustrate. The overuse and misapplication of antibiotics during the last century has undermined their effectiveness today. By 2000 as many as 70 percent of pneumonia samples were found to be resistant to at least one first-line antibiotic, and multi-drug-resistant strains of *Salmonella typhi* (the bacterium that causes cholera and typhoid) have become endemic throughout South America and Africa, as well as many parts of South and East Asia (World Health Organization 2000). Hospital-acquired *Staphylococcus aureus* infections have already become widely resistant to antibiotics, even in the wealthiest countries. Many strains remain susceptible only to the last-resort antibiotic vancomycin, and even this drug is now diminishing in effectiveness (Enright et al. 2002; World Health Organization 2000). The longer we go on applying the same antibiotics, the less effective they become. It takes on average between 12 and 24 years for a new antibiotic to be developed and approved for human use. Pathogens begin developing resistances to these drugs in just a fraction of this time (World Health Organization 2000). So, our current antibiotics will become ineffective over time.

Weaning ourselves from antibiotics requires effective preventive medicine. One such program that has begun gaining popularity among nutritionists, immunologists, and pathologists is probiotics, the practice of cultivating our own natural microbial flora. Microbes live on virtually every external surface of our bodies, including the surface of our gastrointestinal tract and all of our mucous membranes, so they are natural first lines of defense against disease. The cultivation of health-promoting bacterial symbiotes has been shown to enhance the immune system, provide essential nutrition to the host, and decrease the likelihood of colonization by hostile microbes (Bocci 1992; Erickson and Hubbard 2000; Mai and Morris 2004). Even HIV rates have been shown to decrease among people hosting healthy microbial populations (Miller 2000).

Our natural microbial flora has no effect against many diseases, but wherever a colony of pathogenic organisms could thrive, so conceivably could the right innocuous species. So an innocuous microbe could successfully compete against the pathogen and prevent it from thriving in the human environment. Genetic engineering and artificial cell technologies offer two ways we could develop novel probiotics to compete against specific pathogens. These solutions, moreover, would be viable in the long term and more beneficial than present techniques, as a method of curing as well as preventing disease.

However, the precautionary principle would bar us from taking advantage of these new long-term weapons against disease. It is impossible to be certain that a new probiotic would cause no problems in the future, especially when human testing is out of the question (again, because of the precautionary principle). Though releasing new probiotics into human microbial ecosystems is undoubtedly risky and must be done cautiously, it may prove the only way to prevent deadly global epidemics in the future. It is not at all implausible that the consequences of inaction far outweigh the potential risks of these technologies. Excessively narrow and restrictive versions of the precautionary principle leave us trapped like a deer in the headlights.

Society's initial decisions concerning artificial cells will be made in the dark. The potential benefits of artificial cells seem enormous, but so do their potential harms. Without gathering a lot more basic knowledge, we will remain unable to say anything much more precise about those benefits and risks. We will be unable to determine with any confidence even the main alternative kinds of consequences that might ensue, much less the probabilities of their occurrence. Given this lack of concrete knowledge about risks, an optimist would counsel us to pursue the benefits and have confidence that science will handle any negative consequences that arise. The precautionary principle is a reaction against precisely this kind of blind optimism. Where the optimist sees ignorance about risks as opening the door to action, the precautionary thinker sees that same ignorance as closing the door.

As an alternative to the precautionary principle and to traditional risk assessment, we propose dropping the quest for universal ethical principles, and instead cultivating the right virtues we will need for deciding in the dark. Wise decision making will no doubt require balancing a variety of virtues. One obvious virtue is caution. The positive lesson of the precautionary principle is to call attention to this virtue, and its main flaw is to ignore other virtues that could help lead to wise decisions. Another obvious virtue is wisdom; giving proper weight to different kinds of evidence is obviously important when deciding in the dark.

We want to call special attention to a third virtue that is relevant to making proper decisions: *courage*. That is, we advocate carefully but proactively pursuing new courses of action and new technologies when the potential benefits warrant it, even if the risks are uncertain. Deciding and acting virtuously requires more than courage. It also involves the exercise of proper caution; we should pursue new technologies only if we have vigilantly identified and understood the risks involved. But the world is complex and the nature and severity of those risks will remain somewhat uncertain. This is where courage becomes relevant. Uncertainty about outcomes and possible risks should not invariably block action. We should weigh the risks and benefits of various alternative courses of action (including the "action" of doing nothing), and have the courage to make a leap in the dark when on balance that seems most sensible. Not to do so would be cowardly.

We are not saying that courage is an overriding virtue and that other virtues like caution are secondary. Rather, we are saying that courage is one important virtue for deciding in the dark, and it should be given due weight. Precautionary thinking tends to undervalue courage.

Our exhortation to courage is vague, of course; we provide no mechanical algorithm for generating courageous decisions. We do not view this as a criticism of our counsel for courage. For the reasons outlined in the previous section, we think no sensible mechanical algorithm for deciding in the dark exists. If we are right, then responsible and virtuous agents must exercise judgment, which involves adjudicating conflicting principles and weighing competing interests. Because such decisions are deeply context-dependent, any sensible advice will be conditional.

New technologies give us new powers, and these powers make us confront new choices about exercising the powers. The new responsibility to make these choices wisely calls on a variety of virtues, including being courageous when deciding in the dark. We should be prepared to take some risks if the possible benefits are significant enough and the alternatives unattractive enough.

Conclusions

Artificial cells are in our future, and that future could arrive within a decade. By harnessing the automatic regeneration and spontaneous adaptation of life, artificial cells promise society a wide variety of social and economic benefits. But their ability to self-replicate and unpredictably evolve creates unforeseeable risks to human health and the environment. So it behooves us to start thinking through the implications of our impending future with artificial cells now.

From the public discussion on genetic engineering and nanotechnology one can predict the outline of much of the debate that will ensue. One can expect the objections that creating artificial cells is unnatural, that it commodifies life, that it fosters a reductionistic perspective, and that it is playing God; but we have explained why these kinds of considerations are all unpersuasive.

Utilitarianism and decision theory offer scientifically objective and pragmatic methods for deciding what course to chart, but they are inapplicable when deciding in the dark. No algorithm will guarantee sound policies as long as society is largely ignorant of the potential consequences of its actions. The precautionary principle is being increasingly applied to important decisions in the dark, but the principle fails to give due weight to potential benefits lost through inaction (what we called "harms of inaction"). We suggest that appropriately balancing the virtues of courage and caution would preserve the attractions of the precautionary principle while avoiding its weaknesses.

Acknowledgments

This research was supported by a Ruby Grant from Reed College. For helpful discussion, thanks to Steve Arkonovich, Hugo Bedau, Todd Grantham, Scott Jenkins, and Leighton Reed, and to audiences at the LANL-SFI Workshop on Bridging Non-Living and Living Matter (September 2003), at the ECAL'03 Workshop on Artificial Cells (September 2003), at a philosophy colloquium at Reed College (October 2003), and at a philosophy seminar at the College of Charleston (November 2003).

Notes

1. For discussions of this argument as applied to other forms of biotechnology, see Kass 2002, chap. 6; and Comstock 2000, 196–198.

2. See Dobson 1995 and chap. 10 of Kass 2002 for discussions of this objection in other contexts.

3. For discussions of the playing-God objection as it has entered into the genetic engineering controversy, see Comstock 2000, 184–185; and Reiss and Straughan 1996, 79–80, 121.

4. For one attempt to identify the key triggers for stages of ethical action regarding artificial cells, see Bedau et al. 2009.

5. This principle to our knowledge was first discussed in Stich 1978.

6. One notable exception to this pattern is the development of new drugs, which must be proven to be safe and effective before being allowed into the public market.

7. The formulation of the precautionary principle adopted at the Wingspread Conference (Raffensperger and Tickner 1999, 353f) and the ETC Group's formulation (ETC Group 2003, 72) are very similar to the formulation quoted in the text.

8. Appendix B in Raffensperger and Tickner 1999 is a useful compilation of how the precautionary principle is used in international and U.S. legislation.

9. Attempts to defend the precautionary principle and make it applicable in practice are collected in Raffensperger and Tickner 1999; Morris 2000 presents skeptical voices.

References

Bedau, M. A., E. Parke, U. Tangen, and B. Hantsche-Tangen. 2009. "Social and Ethical Checkpoints for Bottom-Up Synthetic Biology, or Protocells." *Systems and Synthetic Biology* 3: 65–75.

Benner, S. A., A. Ricardo, and M. A. Carrigan. 2004. "Is There a Common Chemical Model for Life in the Universe?" *Current Opinion in Chemical Biology* 8: 672–689.

Berg, P., D. Baltimore, H. W. Boyer, et al. 1974. "Potential Biohazards of Recombinant DNA Molecules." *Science* 185: 303.

Bocci, V. 1992. "The Neglected Organ: Bacterial Flora Has a Crucial Immunostimulatory Role." *Perspectives in Biology and Medicine* 35 (2): 251–260.

Brannigan, M. C., ed. 2001. *Ethical Issues in Human Cloning: Cross-Disciplinary Perspectives*. New York: Seven Bridges Press.

Comstock, G. L. 2000. *Vexing Nature? On the Ethical Case against Agricultural Biotechnology*. Boston: Kluwer Academic.

Crichton, M. 2002a. "How Nanotechnology Is Changing Our World." *Parade*, November 24, 6–8.

———. 2002b. *Prey*. New York: HarperCollins.

Dobson, A. 1995. "Biocentrism and Genetic Engineering." *Environmental Values* 4: 227.

Dyson, F. J. 2003. "The Future Needs Us!" *New York Review of Books*, February 13, 11–13.

Enright, M. C., D. A. Robinson, and G. Randle. 2002. "The Evolutionary History of Methicillin-Resistant *Staphylococcus aureus* (MRSA)." *Proceedings of the National Academy of Sciences of the United States of America* 99 (11): 7687–7692.

Erickson, K. L., and N. E. Hubbard. 2000. "Probiotic Immunomodulation in Health and Disease." *Journal of Nutrition* 130 (2): 403S–409S.

ETC Group. 2003. *The Big Down: From Genomes to Atoms*. Accessed June 2003. www.etcgroup.org.

Geiser, K. 1999. "Establishing a General Duty of Precaution in Environmental Protection Policies in the United States: A Proposal." In Raffensperger and Tickner 1999.

Joy, B. 2000. "Why the Future Does Not Need Us." *Wired* 8 (April). Accessed September 2007. www.wired.com/wired/archive/8.04/joy.html.

Kass, L. R. 2002. *Life, Liberty, and the Defense of Dignity: The Challenge for Bioethics*. San Francisco: Encounter Books.

Mai, V., and J. G. Morris, Jr. 2004. "Colonic Bacterial Flora: Changing Understandings in a Molecular Age." *Journal of Nutrition* 134 (2): 459–464.

Merkle, R. 1992. "The Risks of Nanotechnology." In *Nanotechnology Research and Perspectives*, edited by B. Crandall and J. Lewis. Cambridge, MA: MIT Press.

Miller, K. E. 2000. "Can Vaginal Lactobacilli Reduce the Risk of STDs?" *American Family Physician* 61 (10): 3139–3140.

Morris, J., ed. 2000. *Rethinking Risk and the Precautionary Principle*. Oxford: Butterworth-Heinemann.

Pace, N. 2001. "The Universal Nature of Biochemistry." *Proceedings of the National Academy of Sciences* 98: 805–808.

Parke, E. C., and M. A. Bedau. 2009. "The Precautionary Principle and Its Critics." In *The Ethics of Protocells: Moral and Social Implications of Creating Life in the Laboratory*, edited by M. A. Bedau and E. C. Parke. Cambridge, MA: MIT Press.

Raffensperger, C., and J. Tickner, eds. 1999. *Protecting Public Health and the Environment: Implementing the Precautionary Principle*. Washington, DC: Island Press.

Rasmussen, S., M. A. Bedau, L. Chen, et al., eds. 2009. *Protocells: Bridging Nonliving and Living Matter*. Cambridge, MA: MIT Press.

Reiss, M. J., and R. Straughan. 1996. *Improving Nature? The Science and Ethics of Genetic Engineering*. Cambridge: Cambridge University Press.

Resnick, M. 1987. *Choices: An Introduction to Decision Theory*. Minneapolis: University of Minnesota Press.

Ropeik, D., and G. Gray. 2002. *Risk: A Practical Guide to Deciding What's Really Safe and What's Really Dangerous in the World around You*. Boston: Houghton Mifflin.

Silver, L. M. 1998. "Cloning, Ethics, and Religion." *Cambridge Quarterly of Healthcare Ethics* 7: 168–172. Reprinted in Brannigan 2001.

Stephens, T. D., and R. Brynner. 2001. *Dark Remedy: The Impact of Thalidomide and Its Revival as a Vital Medicine*. New York: Perseus.

Stich, S. P. 1978. "The Recombinant DNA Debate." *Philosophy and Public Affairs* 7: 187–205.

Watson, J. D. 2003. *DNA: The Secret of Life*. New York: Knopf.

Wilson, R., and E. A. C. Crouch. 2001. *Risk-Benefit Analysis*. Cambridge, MA: Harvard University Press.

World Health Organization. 2000. *World Health Organization Report on Infectious Diseases 2000: Overcoming Antimicrobial Resistance*. Accessed May 2004. www.who.int/infectious-disease-report/2000/.

CHAPTER SEVEN

Space Exploration and Searches for Extraterrestrial Life

Decision Making and Societal Issues

Margaret S. Race
SETI INSTITUTE

For centuries people have wondered about life and our place in the Universe. Where did life come from and what is its meaning? Are we alone? What is the future of life? Much has been written about the changing historical views of life's origin and the possible existence of extraterrestrial (ET) life (see, for example, Dick 1999, 2002). Today, as we continue to seek answers to these same questions, we are arguably the first generation with sufficient scientific and technological prowess to answer the questions definitively—at least from the scientific perspective. Through systematic research and exploration, we have integrated information across science disciplines and developed a perspective that spans literally from the nanoscale to the cosmos. We may soon be able to determine whether life is uniquely associated with Earth or is instead a cosmic phenomenon, having arisen repeatedly in the Universe whenever and wherever conditions are suitable. The scientific progress is fast-paced, exciting, and energizing, but it also gives us reason to pause and consider the meaning and potential impacts in more than just science realms.

As the scientific quest continues, we are already aware of associated questions that touch on a variety of societal concerns, encompassing ethical, legal, theological, psychological, and other perspectives. What would it mean in theological terms if we verify that extraterrestrial (ET) life exists? What are our ethical and legal responsibilities, if any, toward ET life and environments, and who should be involved in making decisions about research and exploration that might interfere with life elsewhere? Because all inhabitants of Earth must live with the outcome of actions initiated by a scientific and spacefaring minority, it is appropriate to ask whether there are constructive ways for nonscientists, diverse stakeholders, other cultures, and the general public to weigh in on deliberations and decision making about future contact and exploration. As we expand humankind's reach and potential impact on worlds and life beyond Earth, we should acknowledge and avoid the kinds of mistakes made by past generations during their equally ambitious exploration of our world long ago.

Understanding Different Searches for ET Life

Before considering the possible societal questions and issues ahead, it is essential to have a basic overview of the different ways that scientists are searching for evidence of ET life, what they might find, and how they would interpret the data. What is commonly referred to as the *search for extraterrestrial life* is actually a combination of separate research and exploration efforts, using varied technologies and methods, all focused in some way on finding verifiable evidence associated with life beyond Earth. At present, searches for ET life fall into three main categories: (1) The Search for Extraterrestrial Intelligence (SETI), (2) the search for extrasolar and habitable planets, and (3) exobiology and astrobiology. Each of these is described briefly below, and in detail elsewhere (Race 2008).

Searches for ETI seek evidence of life in the Universe by looking for some signature of its technology, mainly by using radio telescopes and signal-processing technology on Earth to detect electromagnetic signals or messages sent across interstellar distances by technological civilizations elsewhere in the Galaxy (see www.seti.org). As Tarter points out in this volume, the search is for the technological manifestation of intelligence elsewhere, not a pure search for intelligent life itself. Detection of a verified signal would be interpreted as evidence for the existence of advanced, intelligent life, but probably would reveal little, if anything, about that life-form's biology, living conditions, motivations, or relationship to life as we know it. In addition, because the signals would have been sent from a galactic location that might be as many as tens to thousands of light-years away, the message would necessarily present information from the past, making it impossible to know the current status or even the continued existence of the message senders.

Searches for extrasolar planets use a variety of remote and telescopic methods to detect and identify planets orbiting other stars beyond our Solar System, with particular interest in finding terrestrial-rocky planets and/or those located in potentially habitable zones of their solar systems. As of 2013, over 3,600 extrasolar planets have been identified using a variety of methods—with more than 850 confirmed, and nearly 2,800 candidates awaiting further examination (see http://planetquest.jpl.nasa.gov). Identifications of extrasolar planets are made by a variety of methods (Doppler shift-stellar wobble; transit method; direct imaging; gravitational microlensing; and so on—see http://planetquest.jpl.nasa.gov/page/methods). Although Jupiter-size planets are easiest to locate, researchers are particularly interested in finding Earth-size planets that orbit Sun-like stars at distances where temperatures are suitable for liquid water to exist. Because all evidence about exoplanets is collected remotely across galactic distances of tens to thousands of light-years, it will be impossible to study the nature of any presumed biotic evidence directly, or to determine the current status or continued existence of biologically interesting features. Even so, the searches are important in providing growing evidence for the existence of locations with conditions and features that may correlate with biotic and/or chemical processes associated with life. In short, searches for extrasolar planets allow us to determine only that a planet may be habitable—but not that it is inhabited.

In contrast, *exobiology and astrobiology research and exploration* include diverse interdisciplinary activities on Earth and within the Solar System that study the origins, evolution, distribution, and fate of life in the Universe, seeking to understand the biological, planetary, and cosmic phenomena associated with life. Astrobiology and exobiology address three fundamental questions: How does life begin and evolve? Is there life elsewhere in the Universe? What is the future of life on Earth and beyond? The overall scope and content of the field is summarized in NASA's Astrobiology Roadmap, which also indicates the linkages between the many science research interests and NASA's space exploration missions (http://astrobiology.nasa.gov/nai).

The key domains of astrobiology exploration include understanding the nature and distribution of habitable environments in the Universe, exploring for habitable environments and life in our Solar System, understanding the emergence of life, determining how early life on Earth interacted and evolved with its changing environment, understanding the evolutionary mechanisms and environmental limits of life, determining the principles that will shape life in the future, and recognizing signatures of life on other worlds and on early Earth (Des Marais et al. 2008). The multi-pronged research efforts incrementally accumulate evidence and information related to life, with the prospect of actually discovering ET life on other planetary bodies. Varied instruments, methods, and disciplinary approaches collect geological, biological, atmospheric, physical, and chemical data as well as remote images and other indirect data indicative of essentially real-time conditions.

Based on extensive research and exploration, it is presumed that any ET life discovered in the Solar System would likely be microbial, and the evidence could be from extant, dormant, dead, or fossilized life-forms or even remnant materials or biomarkers of various sorts. The close distances and relatively short times for retrieval of samples mean it would be possible to examine, verify, and characterize the evidence in laboratories, comparing it to life as we know it, and determining its relation, if any, to life on Earth. Thus, there is the potential for cross-contamination that raises environmental health and safety (EHS) concerns on Earth as well as questions of planetary protection for both Earth and celestial bodies. Neither EHS nor planetary protection concerns are associated with searches beyond our Solar System.

An Approach for Considering Societal Issues Associated with Searches

For all three categories of searches, it is possible even now to identify an assortment of societal challenges. In this context, questions arise in legal, theological, philosophical, ethical, psychological, and assorted other humanities areas (see, for example, Bertka 2009; Race 2007; Billingham et al. 1999; Hargrove 1986; Randolph et al. 1997; Almár 2002; McKay 1990; McKay et al. 1991; Davies 1999; Tough et al. 2000; Peters 1994; Harrison 1997). For the most part, researchers in each of the three search categories have proceeded separately in discussing their particular plans or problems. However, the growing interdisciplinary connections across sciences and search types make it reasonable to analyze the societal issues in all three searches together—if only to identify gaps in information or similar concerns, now or in the future (Race et al. 2012).

As a way to fully understand the mix of issues, it is helpful to focus on three different time frames: first, in the current period of science exploration while we are uncertain that ET life exists; then in the months or early years postdetection when we will likely deliberate about how to respond to a verified discovery of ET life; and finally, in the very long term, extrapolating to scales of decades or centuries, as humankind likely extends beyond Earth in significant ways, perhaps interacting with ET life. In addition to examining issues at different time periods, it is also important to anticipate what decision-making processes and sociopolitical contexts will be used to guide actions and plans—evaluating in advance how quantitative information versus qualitative concerns will be incorporated into deliberations, and considering features such as the locus of authority, responsibility, and legitimacy in the process.

Procedural Roadmap and a "Best-Process" Approach

Obviously there is no single administrative structure or process that can ensure unanimous agreement in all public debates—particularly when deliberations involve new technologies or unusual applications. However, experiences in environmental decision making over the past several decades have fostered development of what amounts to a procedural roadmap for guiding controversial or challenging public debates. Regardless whether a proposal is to clear-cut a forest, construct a nuclear power plant, or launch a space mission, it is possible to identify, understand, and mitigate potentially significant impacts prior to committing large amounts of money or making decisions that may be irreversible. Ideally, deliberations should build on the most accurate, up-to-date science information available; have a clear, acceptable process for reaching decisions; and include ways to incorporate relevant input from experts, stakeholders, and the public.

A recent report from the Nuclear Regulatory Commission (NRC 2009), *Public Participation in Environmental Assessment and Decision Making*, concluded that broadening deliberations to include public participation generally improves the quality of governmental decisions about the environment and increases the legitimacy of decisions in the eyes of the diverse stakeholders affected by them. Although the principles and approaches in the report were developed from the context of U.S. federal environmental laws and agencies, they provide a useful format for examining how facts, people, organizations, values, cultures, methods, and uncertainty might impact complex decision making, especially when scientific and technical experts play strong roles, as they do in searches for ET life.

The NRC report notes there are no specific tools or techniques that constitute "best practices" for all contexts or types of challenges. Instead it advocates a "best-process" approach centered on diagnostic questions to assess the challenges to decision making and public participation. The diagnostic questions help focus attention on important areas such as scientific data, the decision-making process, stakeholders and their access to information and power, and other potential complications. Borrowing heavily from the NRC approach (NRC 2009, box 9.1), a similar set of diagnostic questions was developed for examining societal issues likely to arise in the context of searches for ET Life (see figure 7.1).

Scientific context:

What information is needed for decision-making? What is the degree of availability, adequacy, and uncertainty of data? What is the role of experts in interpreting the basic data and translating it for incorporation into applied contexts? Is there significant polarization of opinion or interpretation?

Methodological Context

Are there generally accepted methods/processes available for dealing with the issues in a public context at appropriate stages in time? (e.g., exploration, post-detection, long term). If not, are there analogues that may be adapted? Are the methods able to involve both expert and non-expert inputs in appropriate ways?

Who's Involved:

Who are the responsible agencies, decision leaders, experts, relevant practitioners, and diverse stakeholders that will have interest in the deliberations and their outcome? Is there clarity about their roles, authorities and mandates? How do values, interests, cultural views and perspectives differ among the participants?

Abilities and Constraints of Participants:

Are all interested and affected stakeholders adequately accommodated in the decision making process, or not (informed, involved, adequately represented, and with power to influence the process in appropriate ways etc.)? Who are the missing stakeholders or interests — and is it possible to include them in some way?

7.1 Diagnostic Questions to Examine the Context of Decision Making and Consideration of Societal Concerns in ET Searches and Exploration

In the sections below, these questions are loosely used to examine how decisions will be made and societal issues aired for the three search types at different time periods. The analysis is not meant to suggest specific recommendations or decisions about future plans, but rather to explore whether and how nontechnical concerns may be incorporated into legitimate deliberation processes. The intent is to identify those situations and issues that can be handled by existing decision-making processes versus those that may require modification or special attention in order to accommodate unusual features associated with space exploration and ET life.

Issues during Exploration: Implications of Ongoing Pursuits

Overall, the activities undertaken in the three search categories during current exploration already use the equivalent of the best-process approach. Each operates as part of science in a democratic system, complete with oversight and review processes established by government agencies, institutions, and professional groups. However, the review and decision-making processes affect the three search types differently.

Current searches for both ETI and extrasolar planets are uncomplicated and raise few, if any, unusual societal questions. Because both search categories use indirect research methods, collect remote data from great distances, and rely on conventional astronomical approaches, neither raises questions that require special regulatory oversight or public scrutiny, except perhaps in the construction of Earth-based facilities and telescopes. All data are available and communicated widely to the scientific community and the public. Any uncertainty or debate about the way searches are undertaken are simply matters of science and technology, and do not raise public concerns about transparency, trust, or stakeholder exclusion.

Astrobiological/exobiological searches within the Solar System involve a mix of direct and indirect methods and real-time findings (as opposed to data from many light-years away), which raise questions about potential contamination, both on Earth and beyond. Research conducted on Earth, whether in laboratories or outdoors, must adhere to relevant regulations, reviews, and permit processes, which range from minor and routine for individual projects to more complicated public reviews and environmental impact analyses for larger projects and launched missions. Even so, for most astrobiological activities, existing decision-making processes are adequate for handling the range of societal questions, which at this time fall mostly into EHS concerns. If anything, the uncertainty about the existence of ET life has added an unusual wrinkle to preparation of risk assessments and anticipation of possible impacts of future plans. To date, only two particular areas have seen extra scrutiny: (1) proposed sample return missions that deliberately return ET materials to Earth, and (2) planetary protection policy under the Outer Space Treaty. In both cases, special processes and frameworks have been developed for dealing with the unusual concerns.

Issues Associated with Sample Return Mission

Discussions about sample return missions (for instance, the Apollo program; Mars missions; sample return from small Solar System bodies) have centered entirely on questions of the science and technology of quarantine and biocontainment, emphasizing biosafety, adequacy of facilities, oversight, and development or handling protocols if needed (see, for example, Allton et al. 1998; Rummel et al. 2002; NRC 1998). If evidence for ET life is found during future testing on Earth, the current Mars protocol indicates that all activities should be halted until a science advisory group reassesses and validates the adequacy of both the containment facility and the protocol test plans. The approach emphasizes science and technical data related to potential biohazards, and also stresses the importance of transparent administrative oversight and public communication. Currently, sample-handling plans do not stipulate whether or how to address societal concerns (such as ethical, cultural, and public perspectives about having ET life on Earth; who besides scientists should be represented in decision making; and how to deal with legal ambiguities).

Planetary Protection Policy: Scientific and Ethical Issues

Perhaps the greatest scrutiny of searches in the Solar System comes in the form of planetary protection policy, which traces back to the Outer Space Treaty of 1967 (UN 1967). It requires that space exploration be done in ways that avoid harmful contamination of planets by either forward contamination of target bodies by hitchhiker microbes from Earth, or backward contamination of Earth by extraterrestrial materials, returning crew, or samples upon return. For the past four decades, the Committee on Space

Research (COSPAR), a nongovernmental international scientific body, has been responsible for formulating and revising policies as needed in response to updated science information. The rationale behind planetary protection policy has been to preserve planetary conditions for future scientific exploration related to biological and chemical evolution (for a review of the history of planetary protection, see NRC 2006, chap. 2). However, in light of significant recent advances in understanding both planetary environments and microbiology, this long-held emphasis on protection for mainly scientific reasons has been challenged.

An NRC study on forward contamination of Mars (NRC 2006) recommended an interdisciplinary analysis to determine whether the framework for planetary protection policy and practices should be extended to include ethical considerations. Subsequent discussions by NASA, COSPAR, and other space law and science organizations led an international workshop in 2010 to discuss whether revisions to planetary protection policies are necessary to address ethical concerns related to planetary environments and possible indigenous ET life (Rummel et al. 2012). The workshop was the first international effort of its kind to officially address possible ethical implications regarding ET life and environments, while maintaining science protection objectives. The workshop also explicitly considered how to involve nonscientists and the public in the dialogue.

Other Issues

Although the workshop initiated important discussions in areas beyond science, there still remain notable unaddressed policy areas. For example, should environmental management or use policies be different on bodies where no ET life exists (such as the Moon and most asteroids) versus those where conditions are amenable to life (such as Mars, Europa)? Rather than taking repeated incremental steps to address issues (see, for example, Laursen 2011; NASA 2010), is there instead a need for an international decision-making process or organization(s) to guide and/or oversee the commercial and private sectors as they develop plans for future activities like exploration, extraction, tourism, and commercialization? Should policies be developed to require biologically reversible exploration until we know whether ET life exists on a body (McKay 2009)? Under what conditions would terraforming be warranted as a way to modify environments on a planetary scale (McKay and Marinova 2001)? Resolving these and other questions will require both scientific analyses and input from numerous outside experts and perspectives. Although discussions have begun to consider ways to address the activities and issues ahead (for example, Ehrenfreund et al. 2012; Race 2011), currently there exist no clear policy or decision-making frameworks to balance diverse stakeholder objectives with scientific data and international policy aims.

Issues after ET Discovery: Operating Rules and Guidelines

In considering questions that may arise after the anticipated discovery of ET life, most discussions about postdiscovery have emphasized primarily how scientists, technical experts, and government agencies will decide what to do next. Again, the response is scenario-dependent.

Extrasolar Planets

In general, the recent discoveries of numerous extrasolar planets raise no unusual societal questions. A successful "discovery" is merely suggestive of possible habitable environments with no current plans or prospects for visiting or returning materials. The absence of public concern about the growing number of discovered exoplanets is indicative of the emphasis on primarily scientific matters rather than societal questions.

SETI Signals and SETI Principles

Although this has not yet occurred, receipt of a signal from ETI would likely pose no immediate concerns about visits or physical exchanges, meaning no additional legal or procedural steps are needed. Even so, the SETI community has already recognized the importance of anticipating how a discovery would be handled, scientifically, technically, internationally, and with the public; as a result it has taken extensive steps to consider its actions postdetection.

A set of SETI Principles for postdetection were approved and adopted in 1989 by the SETI Committee of the International Academy of Astronautics (IAA). Focusing on human response at the time of discovery, the Principles provide step-by-step operational guidelines for verifying the signal, sharing information openly, and consulting broadly and internationally prior to sending any return message (see Billingham et al. 1999 for full text). In the intervening years, the IAA established a permanent SETI

Study Group to organize and discuss topics related to scientific SETI efforts. A special task group on SETI Post-Detection is charged to "prepare, reflect on, manage, advise, and consult in preparation for and upon the discovery of a putative signal of extraterrestrial intelligent (ETI) origin" (see http://iaaseti.org). Even with this organized group in place, it is recognized that various complications remain (recommendations are unenforceable; implementation problems are likely due to lack of organizational readiness; plans are limited for handling mass media communication, false positives, or individuals acting on their own; and so on). Moreover, the SETI Principles explicitly postpone for future deliberations any decisions about postdetection response actions, acknowledging the need to include representatives of "humankind," not just scientists and technical experts, in considering whether and how to respond. In addition to these efforts, there has also been considerable research on many postdetection topics, including psychological and theological impacts of discovery, media communication with the public, and other areas of interest to the public. When social scientists, historians, religious leaders, and diplomats joined SETI researchers to consider various cultural aspects of detection (Billingham et al. 1999), they concluded that short-term reaction to an announcement would likely unfold in accordance to personal, religious, and political belief systems at the time.

ET Life within the Solar System

Despite the fact that an ET discovery could happen at any time, there are currently no official policies or recommendations by either COSPAR (international) or NASA applicable to the discovery of microbial ET life during astrobiology research or exploration—whether it is found in a lab on Earth, in situ by robotic probes, or by astronauts during future missions. Aside from the tentative recommendations to place a temporary halt on analyses of Mars samples in any Earth-based sample-receiving facility (Rummel et al. 2002), the only other clear postdiscovery guidance is to keep samples contained until further characterization studies are completed. No discussions have occurred on whether a discovery of microbial ET life would impact future missions or planetary protection—and in what ways.

Beyond the usual EHS questions and planetary protection concerns, there is no consensus on the legal, ethical, theological, or psychological implications of microbial ET life (Race and Randolph 2002). In the meantime, there are many unresolved questions, some overlapping with SETI concerns, and others touching on new areas. For example, what are the implications of verifying in real time a possible second genesis for life? How would a discovery impact different religious or cultural traditions, and could those different interpretations impact continued scientific work? What would our obligations be toward a life-form different from life as we know it, and would its uniqueness argue for a different ethical approach than that applied to microbes on Earth? Would it be justifiable to undertake additional sample return missions, or to proliferate or sterilize the ET microbes in the context of laboratories studies? It is not clear what decision-making process would be used or how decision leaders would be selected to address these questions, although one could expect that scientists and spacefaring agencies would likely take the lead in deliberations. Who would represent the interests of the ET life-form or its environment (if appropriate), and how would nonscientists, nonspacefaring nations, the public, and diverse cultures be included in the discussions? One can only speculate whether issues of trust and transparency will complicate future deliberations, even though the scientific community and NASA are pledged to communicate openly as part of the existing protocols.

Even as we continue to grapple with environmental ethics on Earth (Rolston 1988), there is uncertainty over what "rights," if any, should be afforded a verified ET living entity and how we would address questions like ownership, patenting, environmental management, or private sector activities on the native planet. Because current decision-making approaches and methods center largely on information about life as we know it, would it be necessary to develop and then shift to a cosmocentric perspective that deliberately goes beyond anthropocentric thinking about life (Lupisella and Logsdon 1997; Lupisella 1998; Dick and Lupisella 2009)? If so, who should decide when and whether such a shift is appropriate? Because answers to questions may have direct impacts on future governmental and private sector activities, they are likely to have more urgency than deliberations about distant, indirect ET discoveries.

Long-Term Postdiscovery: Implications of Human Activities

Looking decades or centuries hence, perhaps the most important questions relate to interactions with ET life and their impacts on life as we know it. In many ways the issues are similar to those being asked about new technologies like synthetic biology and nanotechnology. Each involves breakthroughs in science, follow-on incremental actions, concerns about accidents and adequacy of technological control measures, and questions about the nature of the "new" life and future impacts upon life and environments on Earth, as well as future evolutionary trajectories. Perhaps the discovery of ET life, particularly in the Solar System, will prompt a repeat of the lengthy public debate in the early years of biotechnology and genetic engineering, when there was an abrupt shift from deliberations among mainly scientific experts, to public examination of science and societal issues together in legislative, executive, and judicial arenas and the mass media (Race 2007). Even after the furor subsided and reviews of recombinant DNA research proposals have become routine, procedural and regulatory questions still arise in response to new developments (see, for example, Race and Hammond 2008; Race et al., 2012).

Assuming we continue an incremental approach to decision making about science and technology, it will be important that interested stakeholders and diverse nonscience perspectives are included in deliberations about ET searches, exploration, and discoveries. Without a doubt, it is difficult to anticipate the ramifications of our discoveries and likely expansion beyond Earth, whether exploring, visiting, or establishing settlements. Even so, we should consider the general directions launched by our current policies and guidelines, and ponder how effective they might be judged in hindsight. For example, assuming that the Outer Space Treaty continues in force, it would be instructive to consider what type of governance or regulations would be effective to safeguard special areas or entire planets (see, for example, Cockell and Horneck 2006; Rummel et al., 2012), and whether it would even be appropriate to continue treating Outer Space as a resource for humankind.

While these topics may seem far off or hypothetical today, we owe it to ourselves and future generations to anticipate where we are heading with science and exploration. Unlike generations of centuries past, we are aware that our deliberate actions and technologies have the very real prospect of modifying life's evolutionary trajectory on literally planetary scales and in relatively short times. Even as science and technology describe visions of future possibilities, we should grapple with the broad significance, meaning, and advisability by including input from the public and experts in many nonscience fields. The question is whether we can do a better job of guiding technologies to maximize their benefits while avoiding the kinds of problems created by past unguided progress (pollution and industrialization, nuclear waste disposal, overexploitation of resources, lack of environmental planning, global warming, and so on). Already, astrobiology scientists have joined with experts in humanities and social sciences to explore the issues (Race et al. 2012).

Conclusion

In light of the many ET discovery scenarios possible and the different search types and time periods involved, it may be appropriate to develop a more coordinated approach to analyzing the societal implications ahead. At the very least, those involved should be aware of the importance of crosscutting issues and findings for all search types, not just their own. Moreover, it is important to recognize that current deliberations and decision making are almost exclusively in the realm of scientific and spacefaring elites, whose communications with the public are largely educational and often one-way (except perhaps during environmental impact deliberations). Considering the potential significance of ET life discoveries, it is appropriate to be proactive in expanding the dialogue with experts in nonscience fields, inviting them to become informed participants and information providers in future deliberations.

As we ponder the future together, we should acknowledge that science and technology are embedded inseparably in societal and cultural contexts, and that all perspectives are needed as we deliberate about life's collective future, whether that of humans, Earth's biota, or extraterrestrials. Just as informed consent is understood and practiced in deliberations about new or experimental biomedical treatments, it is appropriate to see if we can figure out the equivalent of "informed societal consent" as we anticipate the implications of space exploration, including the possible discovery of ET life. Now is the time to analyze the issues thoroughly and proactively in order to develop a fuller understanding of the significance of discovery, along with the broad societal implications and possible impacts on life's future.

References

Allton, J., et al. 1998. "Lessons Learned during Apollo Lunar Sample Quarantine and Sample Curation." *Advances in Space Research* 22 (3): 373–382.

Almár, I., 2002. "What Could COSPAR Do for the Protection of the Planetary and Space Environment?" *Advances in Space Research* 30 (2002): 1577–1581.

Bertka, C., ed. 2009. *Exploring the Origin, Extent and Future of Life: Philosophical, Ethical and Theological Perspectives.* Cambridge: Cambridge University Press.

Billingham, J., et al. 1999. *Social Implications of the Detection of an Extraterrestrial Civilization: A Report of the Workshops on the Cultural Aspects of SETI.* Mountain View, CA: SETI Institute.

Cockell, C. S., and G. Horneck. 2006. "Planetary Parks: Formulating a Wilderness Policy for Planetary Bodies." *Space Policy* 22: 256–261.

Davies, P. 1999. *The 5th Miracle: The Search for the Origin and Meaning of Life.* New York: Touchstone.

Des Marais, D. J., et al. 2008. "The Astrobiology Roadmap." *Astrobiology* 8 (4): 715–730.

Dick, S. J. 1999. *Life on Other Worlds: The Twentieth Century Extraterrestrial Life Debate.* Cambridge: Cambridge University Press.

———, ed. 2002. *Many Worlds: The New Universe, Extraterrestrial Life and the Theological Implications.* Philadelphia: Templeton Foundation Press.

Dick, S. J., and M. L. Lupisella, eds. 2009. *Cosmos and Culture: Cultural Evolution in a Cosmic Context.* Document no. SP-2009-4802. Washington, DC: NASA.

Ehrenfreund, P., C. McKay, J. D. Rummel, B. H. Foing, C. R. Neal, T. Masson-Zwaan, M. Ansdell, N. Peter, J. Zarnecki, S. Mackwell, M. A. Perino, L. Billings, J. Mankins, and M. Race. 2012. "Toward a Global Space Exploration Program: A Stepping Stone Approach." *Advances in Space Research* 49: 2–48.

Hargrove, E. C., ed. 1986. *Beyond Spaceship Earth: Environmental Ethics and the Solar System.* San Francisco: Sierra Club Books.

Harrison, A. 1997. *After Contact: The Response to Extraterrestrial Life.* New York: Plenum Press.

Laursen, L. 2011. "NASA to Launch Guidelines to Protect Lunar Artifacts." *Science* 333 (September 2): 1207–1208.

Lupisella, M. 1997. "The Rights of Martians." *Space Policy* 13 (2): 89–94.

———. 1998. "Astrobiology and Cosmocentrism." *Bioastronomy News*, IAU Commission 51m 10 (1): 1–2, 8. Pasadena: The Planetary Society.

Lupisella, M., and J. Logsdon. 1997. "Do We Need a Cosmocentric Ethic?" International Astronautical Federation Congress (Turin, Italy), Paper IAA-97-IAA.9.2.09, 3–5 Rue Mario-Nikis, 75015 Paris.

McKay, C. P. 1990. "Does Mars Have Rights? An Approach to the Environmental Ethics of Planetary Engineering." In *Moral Expertise: Studies in Practical and Professional Ethics*, edited by C. D. MacNiven. New York: Routledge.

———. 2009. "Biologically Reversible Exploration." Policy Forum, *Science* 323 (February 6): 718.

McKay, C. P., and M. Marinova. 2001. "The Physics, Biology, and Environmental Ethics of Making Mars Habitable." *Astrobiology* 1: 89–109,

McKay, C.P., O. B. Toon, and J. F. Kasting. 1991. "Making Mars Habitable." *Nature* 352: 489–496.

NASA. 2010. "NASA's Recommendations to Space-Faring Entities: How to Protect and Preserve the Historic and Scientific Value of US Government Lunar Artifacts." Accessed October 25, 2012. www.collectspace.com/news/NASA-USG_lunar _ historic_sites.pdf.

National Research Council (NRC). 1998. *Evaluating the Biological Potential in Samples Returned from Planetary Satellites and Small Solar System Bodies: Framework for Decision Making*. Washington, DC: National Academies Press. Available at www.nap.edu.

———. 2006. *Preventing the Forward Contamination of Mars*. Washington, DC: National Academy Press. Available at www.nap.edu.

———. 2009. *Public Participation in Environmental Assessment and Decision Making*. Washington, DC: National Academies Press. Available at www.nap.edu.

Peters, T. 1994. "Exotheology: Speculations on Extraterrestrial Life." *CTNS Bulletin* 143, no. 3 (Summer): 1–9.

Race, M. S. 2007. "Societal and Ethical Dimensions of Astrobiology and the Search for Extraterrestrial Life, Chapter 24." In *Planets and Life: The Emerging Science of Astrobiology*, edited by W.R. Sullivan and J. Baross. London: Cambridge University Press.

———. 2008. "Communicating about the Discovery of Extraterrestrial Life: Different Searches, Different Issues." *Acta Astronautica* 62: 71–78.

———. 2011. "Policies for Scientific Exploration and Environmental Protection: Comparison of the Antarctic and Outer Space Treaties." In *Science Diplomacy: Antarctica, Science and the Governance of International Spaces*, edited by P. Berkman et al. Washington, DC: Smithsonian Institution Scholarly Press.

———. 2012. "Astrobiology and Society: Building an Interdisciplinary Research Community." *Astrobiology* 12 (10): 958–965.

Race, M. S., and E. Hammond. 2008. "An Evaluation of the Role and Effectiveness of Institutional Biosafety Committees (IBCs) in Providing Oversight and Security at Biocontainment Labs." *Biosecurity and Bioterrorism* 6 (1): 19–35.

Race, M.S., K. Denning, C. Bertka, S. Dick, A. Harrison, C. Impey, R. Mancinelli, and Workshop Participants, 2012. "Astrobiology And Society: Building An Interdisciplinary Research Community." *Astrobiology*. October 2012, 12 (10): 958–965.

Race, M. S., J. Moses, C. McKay, and K. J. Venkateswaran. 2012. "Synthetic Biology in Space: Considering the Broad Societal and Ethical Implications." *International Journal of Astrobiology* 11: 133–139

Race, M. S.. and R. Randolph. 2002. "The Need for Operating Guidelines and a Decision-Making Framework Applicable to the Discovery of Non-Intelligent Extraterrestrial Life." *Advances in Space Research* 30 (6): 1583–1591.

Randolph, R., et al. 1997. "Reconsidering the Ethical and Theological Implications of Extraterrestrial Life." *Center for Theology and Natural Sciences Bulletin* 17 (3): 1–8.

Rolston, H. 1988. *Environmental Ethics: Duties to and Values in the Natural World*. Philadelphia: Temple University Press.

Rummel, J., M. Race, and G. Horneck, eds. 2012. *COSPAR Workshop on Ethical Considerations for Planetary Protection in Space Exploration*. Paris: COSPAR. http://cosparhq.cnes.fr/Scistr/PPP Reports/PPP_Workshop Report_Ethical Consid erations.pdf.

Rummel, J. D., et al. 2012. "Ethical Considerations for Planetary Protection in Space Exploration: A Workshop." *Astrobiology* 12(11): 1017–1023.

Rummel, J. D., et al. 2002. *A Draft Test Protocol for Detecting Possible Biohazards in Martian Samples Returned to Earth*. Document no. CP-2002-211842. Washington, DC: NASA.

Tough, A., et al. 2000. *When SETI Succeeds: The Impact of High-Information Contact*. Bellevue, WA: Foundation for the Future.

United Nations. 1967. *United Nations Treaty on Principles Governing the Activities of States in the Exploration and Use of Outer Space, Including the Moon and Other Celestial Bodies*. U.N. Doc A/Res/2222 (XXI); TIAS #6347. New York: United Nations.

CHAPTER EIGHT

Astrobiology and Society

The Long View

Christopher P. McKay
NASA AMES RESEARCH CENTER

The current field of astrobiology includes a fundamentally new question. Whereas previous lines of inquiry focused on the origin and distribution of life in the Universe, astrobiology as defined by NASA explicitly includes the question: What is the future of life in the Universe? In the near term this third question of astrobiology relates to the possibility of survival of life from Earth in space, on Mars, and elsewhere in the Solar System. However, it is interesting to consider the long view and the ultimate goal for human action with respect to life in the Universe.

Unlike studies of the past, considerations of the future must include human decisions. The future of life in the Universe depends on our choices and actions. Human choices, while they may be informed by science, depend on ethical, economic, and broad societal considerations.

In this chapter I suggest that the long-term goal for astrobiology should be to enhance the richness and diversity of life in the Universe. I then consider the implications of this goal in terms of our present activities and, in particular, the search for a second genesis of life on Mars through robotic and human exploration.

Enhance the Richness and Diversity of Life in the Universe

We can justify the search for the origin of life on Earth and the search for a second genesis of life on Mars based purely on scientific curiosity. As with all fundamental research, there is no need to define a specific goal to motivate this search. However, the situation is different when we consider the future because the research questions merge with the choices we make. If we decide to determine whether life, including humans, can survive on Mars, and if we decide to implant life on Mars, there is an explicit purpose to these actions and a goal—even if unstated. Research activities are involved, but they would not exist independently of the goal. Possible goals include economic benefit for corporations or individuals, expansion of the human species, the establishment of a second branch of humanity as insurance against worldwide disasters, and so on.

I suggest that the goal for astrobiology should reflect the central importance of the phenomenon of life and the ability of humans to understand, study, and guide that phenomenon. Life is not the only source of value, but it is perhaps the most important source. Some natural lifeless landscapes and features (such as the rings of Saturn) may have value as well due to aesthetic or cultural aspects and thus should be preserved. However, life is unique in that not only can its value be preserved, it can be spread. If life has value, then humans can create value and spread value as they spread life. In this sense a focus on enhancing life connects to the deeply seated motivation for humans to be active creators, or co-creators, helping shape and guide the existing creation.

Following this logic, I suggest that the long-term goal for astrobiology and society be to enhance the richness and diversity of life in the Universe. Adopting this goal implies that we act in three areas. First, we work to maintain the richness and diversity of life on Earth. Second, we search for, and support, a second genesis of life on other worlds. Third, we expand life beyond Earth, as appropriate (for balanced critique of this third goal, see Sullivan's chapter in this volume). In this chapter I do not dwell on the need for environmental stewardship of the Earth. This topic has been extensively considered. It is mentioned here to emphasize its inclusion in an overall view of the future of life. I focus in this chapter on the issues related with life on other worlds and expanding life from Earth.

Second Genesis

On Earth we have discovered only one example of life. When we speak of diversity, we refer to different species within the context of a single unity of genetics and biochemistry. The possibility exists that we will discover on Mars an example of life that differs at such a fundamental level that it would have to represent an origin of life independent of Earth life. The discovery of a second genesis would be of profound philosophical and practical interest.

The presence of two independent origins of life in our Solar System would give strong support to the notion that life is common in the Universe, thereby contributing to ancient and deep questions of origins and helping us understand the framework within which humankind and life relate to the rest of the Universe.

On the practical side, the discovery of another type of life would create the new field of comparative biochemistry. Currently all our knowledge about life and all fields of science that derive from the study of life are based on one example. All life on Earth shares a common set of biomolecules, a common genetic code, and a common ancestor. It is probable that there are many questions about genetics and biochemistry that will not be answered until we have another example of life to compare with Earth life. The ability to compare biochemistries could illuminate fundamental aspects of such fields of study as agriculture, medicine, and cellular bioengineering—with possibly enormous practical implications.

There exists a range of arguments for why we should preserve a second genesis. As I will argue here, the fundamental reason should be the ethical principles related to the value of life and the value of diversity in life. A second genesis adds an entirely new dimension to the diversity of life—even if the sole representatives of that second genesis are microscopic.

However, even setting aside ethical principles, it is still the case that the utilitarian benefit that comes from direct study of a second genesis should also motivate its preservation. The practical benefits of the scientific study of a second type of life and the ecological understanding that would come from assisting in the restoration of that life to a globally distributed ecology would be valuable to humanity in many ways. Direct benefits in medicine and biotechnology would be added to knowledge gained at the levels of the ecosystem and the biosphere, and this would help in the challenge of biosphere management on Earth. Far from being a replacement for an abused biosphere, a reconstructed Mars could be a lesson in how to maintain Earth's biosphere.

It is possible that we will find evidence of past life on Mars with no indication of current organisms. Smith and McKay (2005) have suggested that we might find the frozen but dead remains of ancient life deep in the ice-rich polar deposits. Such remains would allow us to determine if that life represented a second genesis, even if the organisms are long dead from crustal radiation (Smith and McKay 2005). We should still consider the possibility that we could revive the Martian life-forms from the fragments preserved in these ancient frozen sediments. The task might be considerable, and made more so if the genetics and biochemistry are alien, but the knowledge gained would be enormous and undoubtedly of practical value. Aiding a neighboring world to resurrect its genome and restore its biosphere would be a suitable challenge for an advanced spacefaring civilization with the goal of enhancing the richness and diversity of life in the Universe.

Biologically Reversible Exploration

The state of life on Mars, past and present, is still uncertain. It is useful to consider three possibilities: (1) Mars never had life; (2) Mars had, or has, life that shares a common origin with life on Earth; and (3) Mars has life that represents a second genesis.

The situation is the simplest for the first case—no life. In this case our concern would be for the future biological potential of Mars. If we could determine that Mars was not on a path to developing life on its own—an admittedly difficult determination to make—then ecological alterations of Mars could proceed with the aim of making that planet habitable to the widest possible range of life from Earth.

From an ethical perspective the situation is not too complex for the second case: life with a common origin. In this case, there are no ethical barriers to mingling Earth life with Martian life, although there may be practical concerns. Ecosystems developed on Mars become extensions of ecosystems present on Earth.

The case in which Martian life represents a second genesis of life is the most complex and interesting of the three possibilities. I would advocate that if we discover such life, or the fragmented remains of such life, we should immediately alter plans for the exploration of Mars. In particular we should focus our efforts on the resurrection of the Martian life as necessary and the restructuring of the Martian environment so as to allow that life to expand to produce a global biosphere. Contamination from Earth life would have to be rigorously avoided. (From a planetary protection point of view, the contamination of Earth with Martian life would also have to be rigorously avoided. These issues are dealt with in more detail in the chapters in this volume by Race and Sullivan.) The future exchange of life between Earth and Mars that might result from large impacts would also have to be controlled—preventing large impacts is a likely goal for reasons of human safety anyway.

This view is in contrast to suggestions for a noninterference policy toward life on Mars. My approach can be summarized as "Do no harm and provide help as needed." This ethical choice follows from accepting the principle that we (humans) should seek to advance the richness and diversity of life in the Universe. As discussed above, it also follows from a utilitarian perspective that only recognizes human interests. (See Bedau, in this volume, for closely related ethical discussions.) The practical difficulty in applying this view is that we must first determine what constitutes "help" and whether it is "needed." Thus I would aim the focus of scientific research on Mars, following the discovery of a second genesis of life on that planet, toward understanding the biology, ecology, and evolutionary potential of that life. This is a tall order, but one worth undertaking both for the scientific return and in support of the ethical perspective on the richness and diversity of life.

Because we do not know if there is or was life on Mars or if such life is a second genesis, the precautionary principle might lead us to suggest that we explore Mars in a purely sterile manner until we resolve the question of life. (For further discussion of the precautionary principle, see Bedau's chapter in this volume.) However, sterile exploration of Mars is not possible, for several reasons. First and most directly, we have already contaminated the surface of Mars with Earth organisms. After the Viking missions to Mars revealed the harsh surface conditions, the international regulations related to planetary protection were modified (NASA 2005). Spacecraft sterilization was no longer required, and on July 4, 1997, the *Mars Pathfinder* landed on Mars with an estimated 100,000 viable microorganisms (see, for instance, Barengoltz 2005). Similar levels of microorganisms were also carried by the failed *Mars Polar Lander*, the *Spirit* and *Opportunity* rovers, the failed *Beagle* 2 lander, and the *Phoenix* lander (Barengoltz 2005; Salinas et al. 2007).

Secondly, the cost for sterile exploration is high, and many argue that the low probability of life does not justify this cost. In addition, sterile exploration precludes human exploration, and it is likely that human exploration is needed to adequately resolve the question of life. Fortunately there is a way out of this dilemma—an approach I have termed "biologically reversible exploration" (McKay 2007, 2009).

That we have already contaminated Mars with organisms from Earth is certain, but the contamination is limited and contained. Any microbes released into the Martian environment would be quickly killed by the strong biocidal solar ultraviolet radiation. The only organisms that would survive would be those inside the vehicles protected from sunlight. Without water these organisms would remain in a dormant but viable state until the slow accumulation of cosmic radiation would kill them over a few hundred thousand years or more. So Mars has been contaminated by microorganisms from Earth, but that contamination has not spread. It could be removed easily by visiting all of the landing and crash sites and removing all the pieces. This would be a large task, but it would be possible. Thus, the contamination due to the exploration of Mars to date is biologically reversible. If we can maintain this approach, we can preserve the option of acting to restore a second genesis of life on Mars if such life is discovered in the course of future exploration.

With human exploration, sterilization is not an option. Nor is it realistic to imagine that a human base could be so carefully engineered that it would release no microorganisms into the environment. Human missions to Mars will add significantly to the contamination load already on Mars. However, a human base on the surface of Mars is also biologically reversible.

Robotic or human exploration is biologically reversible if it is possible to remove all microbial contamination resulting from this exploration at a cost that is linearly related to the cost of the initial missions.

In situations where contamination grows rapidly from an initial infection, the costs of decontamination are exponentially related to the initial contamination. The spread of rabbits in Australia is a good example. However, imagine that when a breeding pair of rabbits was released in Australia, another pair was released in Antarctica. Now, years later, the decontamination of Australia is exponentially difficult. But the decontamination in Antarctica would be only linearly difficult—we would find the two dead rabbits not far from where they were released, because the Antarctic environment is not conducive to the reproduction of rabbits.

The spacecraft that have landed on Mars to date have all been surface missions. Even at the crash sites, debris from Earth is no more than a few meters into the surface. To reverse the contamination on Mars involves recovering the spacecraft parts from each of these sites and exposing any contaminated dirt to the sterilizing ultraviolet sunlight.

Things become quite different below the surface, away from the biocidal effects of the solar radiation. For example, if robotic or human explorers drill to investigate a subsurface aquifer, any microorganisms introduced could find conditions suitable for growth and the contamination would spread. Thus, biologically reversible exploration would require rigorous sterilization of any components that are involved in subsurface exploration or that go down the drill hole to an aquifer. Similarly, if human explorers establish bases inside caves, as has been suggested, the naturally sterilizing effect of the surface ultraviolet radiation would be lost and contamination would be persistent. Reversing such contamination would be very difficult compared to reversing the contamination of surface operations.

Our exploration of Mars so far has been biologically reversible. Even the contamination associated with human bases on the surface can be biologically reversible. For future missions, sterilization or even cleaning would not be a requirement, but we would need to track all landing sites and pieces on Mars. In addition we would design missions, particularly rovers, to maximize the efficacy of sterilization on Mars by solar ultraviolet radiation. There would be no necessity to even reduce the biological load on spacecraft. It is likely that following this approach will reduce the costs compared to the current method of reducing the biological load but not sterilizing.

Life from Earth

The one place we know has life is Earth. Thus, even as we search for evidence of other life beyond the Earth, we should investigate the possibility that life from Earth can (and should) spread to other worlds (see Sullivan's chapter in this volume for an alternative point of view). The first step for life beyond the Earth is probably Mars. Thus, an important next step is to determine if life from Earth can grow on Mars.

McKay (2004) and others have suggested that a goal for a near-term mission could be a plant growth module. In this experiment the Martian soil would be washed and augmented with nutrients as necessary and then placed into a container on board the robotic lander. Then seeds would be added and the container would be pressurized with Martian air and sealed. The seeds would germinate and grow, eventually producing new seeds. The cycle of growth would continue in the sealed container as long as practical.

Such a growth experiment would provide an organism-level test of soil biohazard, environment, radiation, and Martian gravity. It would also provide a technical and programmatic basis for advanced plant-based life-support systems. In addition it provides a wonderful opportunity for public involvement as well as being symbolic as the first organism to grow, live, and die on another world. In terms of future missions, the growth of a plant would help defuse back-contamination issues for sample return and human missions by directly demonstrating that the Martian soil is not dangerous to Earth life. Along these lines, a plant growth experiment would be a biological precursor to human exploration. A plant growth experiment could be done on Mars in a way consistent with planetary protection with no inadvertent forward contamination of Mars. This is possible because plants can be prepared and grown completely free of bacteria. While Mars would be the ultimate target, a preliminary experiment could be demonstrated on the Moon.

Conclusion

Astrobiology is the study of the origin of life, the distribution of life in the Universe, and the future of life in the Universe. The question of the future of life in the Universe involves science but also involves human choice at a basic level. Thus it is important for us to consider what purposes and values will guide our choices. I suggest that the long-term goal for astrobiology be to enhance the richness and diversity of life in the Universe.

Clearly this begins with maintaining the richness and diversity of life on Earth. Next comes the search for life beyond Earth to give us an understanding of the natural diversity of life. Third, we can investigate the potential for life from Earth to spread beyond.

Discussions of long-term goals may be abstract and distant, but the perspective outlined here has an immediate implication for planetary exploration. The logic of enhancing the richness and diversity of life implies that we change the basis of planetary protection from preserving science to biological reversibility. This is not much of a change for current robotic missions. It also provides a basis extendable to human exploration. Most importantly, biologically reversible exploration preserves, at reasonable cost, future options related to restoration ecology on Mars for an indigenous biota. Thus I recommend that the concept of biologically reversible exploration needs to be part of our mission design, for both robotic and human exploration, even as we explore Mars to determine whether it can be made habitable for life from Earth.

References

Barengoltz, J. 2005. "A Review of the Approach of NASA Projects to Planetary Protection Compliance." IEEE Aerospace Conference. IEEEAC paper no. 1485.

McKay, C. P. 2004. "The Case for Plants on Mars." *Acta Hort. (ISHS)* 642: 187–192.

———. 2007. "Hard Life for Microbes and Humans on the Red Planet." *Ad Astra* 19 (3): 30–31.

——. 2009. "Biologically Reversible Exploration." *Science* 323: 718.

NASA. 2005. "Planetary Protection Provisions for Robotic Extraterrestrial Missions." Document no. NPR 8020.12C. Washington, DC: NASA.

Salinas, Y., W. Zimmerman, E. Kulczycki, S. Chung, and T. Cholakian. 2007. *Bio-Barriers: Preventing Forward Contamination and Protecting Planetary Astrobiology Instruments*. IEEE Aerospace Conference. IEEEAC Paper no. 1216. doi: 10.1109/AERO.2007.352740.

Smith, H. D. and C. P. McKay. 2005. "Drilling in Ancient Permafrost on Mars for Evidence of a Second Genesis of Life." *Planet. Space Sci.* 53: 1302–1308.

CHAPTER NINE

Planetocentric Ethics

Principles for Exploring a Solar System That May Contain Extraterrestrial Microbial Life

Woodruff T. Sullivan III
DEPARTMENT OF ASTRONOMY AND ASTROBIOLOGY PROGRAM, UNIVERSITY OF WASHINGTON

Two things fill the mind with ever new and increasing admiration and awe, the oftener and the more steadily we reflect on them: the starry heavens above and the moral law within.

—KANT, *CRITIQUE OF PRACTICAL REASON*, 1788

The heavens and the philosophy of ethics have been connected for a long time. But the Space Age has upped the ante with its many unprecedented ethical questions arising from the novelty of our actually being able to visit, not just observe, the extraterrestrial space environment. In this chapter I will discuss the ethical issues raised by exploration of planetary bodies and by the possible presence of extraterrestrial life in our Solar System.[1] I will argue for an extension and adaptation of a rigorous environmental ethics stance that has been proposed for Earth. On this scheme, planets have intrinsic value, and especially so if they might potentially harbor life.[2] Such a "planetocentric ethics" treats all planets somewhat as we treat designated wilderness areas on Earth—that is, with a "hands off" approach unless other treatment is strictly justified for scientific or other needs. This system of ethics is the antithesis of one that allows terraforming and other activities that modify environments (see discussion of this alternative in McKay's contribution in this volume).

An analogy has often been made between twenty-first-century exploration of space and seventeenth- and eighteenth-century Europeans probing the New World. The Europeans encountered vast, seemingly endless lands and over the centuries "tamed" them, though in the process they also eventually despoiled huge tracts. We today have the opportunity and duty to learn from these past mistakes on our home planet and now to get it right for our "New World," the rest of our Solar System.

Environmental Ethics in Space

The worldwide space community has dealt with ethical issues in space since the 1960s, but only in a limited way with regard to the issues of this chapter (see also the contributions by Race and McKay in this volume). The first milestone was the Outer Space Treaty of 1967, which prohibits various activities in space (such as use of nuclear weapons, claiming of territory, "harmful contamination") but is often ambiguous and has only weak enforcement provisions. A stronger version of this, usually referred to as the 1979 Moon Treaty (although it would apply to all planetary bodies), has unfortunately remained unratified by the major spacefaring nations.

The fields of exobiology and astrobiology have spurred many further ethical discussions. Astrobiology's inquiries into the origin of life, its history, the very nature of life, and the possibilities of extraterrestrial life raise compelling, fundamental questions. If microbes were to be found on Mars or Europa, it would be of great philosophical, theological, and scientific importance, especially if their biochemistry was different from that of Earth life. With a second example of life, the new datum would imply a Universe possibly rich with life, and we could begin to investigate the nature of life as a phenomenon.

Chris McKay (1990, 2007, and this volume) has been a pioneer in urging that ethical principles be discussed and applied regarding potential life-bearing sites, in particular Mars.[3] He has argued that if technically feasible, the greatest good that we could do with any ecosystem that we found on Mars would be to modify Mars as needed in order to help it reach its greatest potential (more biomass, more biodiversity). He calls this "ecosynthesis" or "restoration ecology," based on the idea that any ecosystem on Mars today is probably a faint remnant of the Martian ecosystem 4 gigayears (Gyr; 1 gigayear equals 10^9 years) ago when Mars was warm and wet and clement, perhaps with a flourishing biosphere. Similarly, McKay argues for an ethical scheme in which a life-filled planet is more valuable than a "dead" one, so that if possible, we ought to bring any lifeless planet to a state where it can support Earth life (terraforming).

In response to missions with astrobiological goals or ramifications, NASA has developed policies on "planetary protection," meaning keeping planets biologically isolated despite interplanetary traffic by machines and humans (each accompanied by $\sim 10^{14}$ microbes). Such policies must be informed by ethical principles as to how we should treat other potentially life-bearing planets, and they must grapple with many technical issues—containment facilities, sterilization of spacecraft, and so on (Rummel 2007; Race 2007, this volume).

NASA has also periodically produced planning documents that emphasize the importance of ethical concerns in astrobiology. For instance, NASA's latest Astrobiology Roadmap (Des Marais et al. 2008) has four basic principles, one of which refers to environmental ethical issues:

"Astrobiology encourages planetary stewardship through an emphasis on protection against forward and back biological contamination and recognition of ethical issues associated with exploration" (17).

In sum, astrobiology and its ethical issues are important today for many space activities beyond Earth orbit. What, then, should be at the core of our ethical principles dealing with the exploration of and possible presence of life on other planets?

Moral Standing of Nonhumans

Environmental ethics is a relatively new branch of philosophy, arising in the mid-twentieth century as civilization's activities began to noticeably impinge on the health of the natural world. Whereas previously ethics had been confined to interactions among humans, some philosophers now began considering the moral standing of animals, plants, ecosystems, microbes, and even inanimate rocks. Humans are always the *moral agents* who make ethical decisions, but who or what else is worthy to be called a *moral subject*?[4] And even if something is a moral subject, how ought one to treat it? Below I give a quick tour of major philosophical ideas on these matters and comment on their applicability and/or desirability in a Solar System context.

Aldo Leopold in his *Sand County Almanac* (1949) argued for a *land ethic* that gives moral standing to an ecosystem based on its evolutionary kinship to us: "A thing is right when it tends to preserve the integrity, stability, and beauty of the biotic community. It is wrong if it tends to do otherwise." This has become a guiding principle in (Earth-bound) environmental ethics, but because Leopold's land ethic is really an ecosystem ethic, its application to lifeless planets is problematic.

Race and Randolph (2002) have discussed one approach for treating any discovered planetary ecosystem and its possible effects on Earth's life. They propose two key principles that are extensions of the Hippocratic oath ("Do no harm") and the golden rule:

1. Cause no harm for planet Earth, its life, and its diverse ecosystems.
2. Respect and do not substantively or irreparably alter the extraterrestrial ecosystem.

Race and Randolph ask us to consider how *we* (and our biosphere) would want to be treated by any superior beings who discovered us, and then to act similarly toward other ecosystems. Such a perspective is indeed illustrative when developing moral guidelines.[5]

In the 1970s Arne Naess set forth a radical environmental philosophy called deep ecology. Naess argues that *all* nonhuman life has intrinsic value, independent of its value to humans for life support, resource utilization, genetic diversity, scientific study, aesthetic appreciation, or spiritual inspiration. These are its key principles (Naess and Sessions 1995):

1. The well-being and flourishing of human and nonhuman Life on Earth have intrinsic value. . . . All have an equal right to live and blossom.
2. Richness and diversity of life forms contribute to the realization of these values and are also values in themselves.

"Life" in the first statement above is taken to include inanimate parts of ecosystems, such as landscapes and rivers. Humans need to be appreciators of all life, rather than manipulators and managers. The core principles are to be violated only to satisfy essential human needs.

Deep ecology is not entirely new, especially if one includes Asian philosophies such as Jainism, an offshoot of Hinduism that teaches that all living things must help each other. In Jainism no living thing, even insects or microbes, should be killed with intent, except to satisfy essential human needs. In the West the essence of deep ecology can be found in the seventeenth-century Dutch philosopher Baruch Spinoza, who saw human matters as secondary to the workings of an infinite, ordered system. This was (and is) a major shift from the usual Judeo-Christian concern with the importance of each person in God's Universe. (For more thorough discussions of Western and Eastern philosophies in astrobiology, see the contributions by Cleland, Stoeger, Hewlett, and Irudayadason in this volume.)

Unlike most "environmentalism" as practiced today, deep ecology says that technical solutions are usually misguided, like giving another drug to a worsening patient already taking a suite of drugs, many already of questionable value and harmfully interacting. Rather than constantly trying to fix ecodisasters created by poor choices, we must fundamentally reorient our activities in this planet's biosphere so as to play only a collaborative role (albeit an important one).

Callicott (1986) has pointed out an important and felicitous consequence when deep ecology is applied to the Solar System. On Earth humans are enmeshed in the web of life and of course cannot even exist without harming many other species and individual organisms. For extraterrestrial life, however, especially if it is of an exotic nature, we can happily avoid harming every one of these species because none are necessary for our own survival.

Finally, Rolston (1986, 1988) has proposed an attractive scheme wherein deep ecology principles are adapted to the Solar System at large. In the next section I will argue that a similar ethical substrate should guide our exploratory activities on planets. Rolston ascribes intrinsic value not just to ecosystems, nor even just to those planets that have potential for ecosystems to be discovered, but to *all* planets. He looks upon planets as "inventive projects" of the natural world with their own natural histories (sequences of projects) that actualize their potentials mostly unknown on Earth. Those locales that seem completely inhospitable, such as Uranus, should be looked upon, not as projects "gone bad," but as part of the quantitative and qualitative diversity of the entire Solar System that led to the arising of Earth's life (and perhaps other Solar System life). An alive Earth would not emerge without all of these projects acting as an ensemble.

We understand only a few of the intimate, deep connections over the past 4.6 Gyr among the Solar System's planets, moons, asteroids, and comets, but these give merit to this philosophical approach. For example, comets and asteroids colliding with Earth have profoundly influenced the history of life on Earth and perhaps even its origin (involving issues such as sterilizations caused by the Late Heavy Bombardment, delivery of organics and volatiles, and mass extinctions). Another fascinating example is the Moon's influence on the history of Earth's life and once again maybe even on its origin (strong early tides, stability of axis tilt and therefore climate). We err if we consider the history and habitability of a planet in isolation from the rest of its planetary system (and its central star). Our presence today on the third planet from the Sun is a product of the *entire* Solar System over its *entire* 4.6 Gyr history. As the naturalist John Muir put it in 1911 in *My First Summer in the Sierra*: "When we try to pick out anything by itself, we find it hitched to everything else in the Universe."

Note that the intrinsic values of Rolston's "projects" do *not* rest on their being viewed or discovered by humans; each has an independent integrity tied into its natural history and its cosmic environment. For instance, consider the Earth's biosphere as it was 2 Gyr ago; that project had no less value because of the lack of humans to observe it or participate in it. Similarly, any microbial life that might exist, say, beneath the icy surface of Saturn's satellite Enceladus is a project with inherent value, independent of its potential discovery by us. Or consider Saturn's moon Titan, which with high probability does not host life today despite its complex organic chemistry at a temperature of ~90 K (–297.67°F or 183.15°C). But it will be a very different project in the distant future when ~5 Gyr from now it may become habitable as the Sun reaches a vastly greater luminosity in its red giant phase (Lorenz et al. 1997). The changing planetary diversity of our Solar System is analogous to the evolving biodiversity that we experience on Earth; every species has intrinsic value as a part of the whole, no matter what our present (imperfect) judgment regarding its importance in the biosphere. Each species is a "project," a product of the inventiveness of the natural world.

Rolston goes further and asks whether or not science itself has intrinsic value—that is, a worth independent of its utility to society in satisfying curiosity, supplying origin stories, manipulating nature, and so forth. If one answers yes, then science's subject, the natural world, must also have intrinsic value, and should remain untrammeled to the greatest extent possible, in order to foster valid scientific research.

Planetocentric Ethics

I propose that the astrobiology community should work to put in force an ethical scheme closely related to Rolston's, one that I call *planetocentric ethics*. In this scheme we extend the concept of *eco-logy* ("home-study" etymologically) such that "home" refers to more than just our neighborhood, or our watershed, or even our biosphere, but indeed refers to the entire Solar System. In planetocentric ethics each planetary body has intrinsic value as part of an ecosystem writ large, and can be modified only with strong justification. In this context, the phrase "planetary body" refers to any substantive physical object, so it can include moons, asteroids, and comets as well as an object meeting the astronomer's definition of planet.

Of course in this or any other system, conflicts will arise between competing values—for instance, between the desire to do scientific research despite some chance of causing significant harm or other change to a planet. A key point of the proposed planetocentric ethics is that the intrinsic value of the planetary body has primacy, and the burden of proof is on the proposed "user" to demonstrate lack of potential harm or significant change. In other words, any proposed activity affecting a planet is presumed "guilty" until proven "innocent" by the preponderance of evidence brought forth. Although a stringent standard, it is ethically sound. If enacted now, such a standard in future centuries, when our interplanetary activities become ever larger and potentially harmful, will preserve the Solar System that we take for granted today.

This is the guiding principle of planetocentric ethics:

> Cause neither physical nor biological harm to any planetary body and its ecosystems.

Regulations that enforce this principle will need to have teeth, perhaps modeled after the 1964 U.S. Wilderness Act, which describes wilderness as "an area where the Earth and its community of life are untrammeled by man, where man himself is a visitor who does not remain." On this criterion virtually the entire Solar System qualifies, *so far*. Possible exceptions are the 1969–1972 *Apollo* landing sites on the Moon, various landing sites on Mars, and Comet Tempel, which was smashed with a several-hundred-kilogram hunk of copper by the *Deep Impact* space probe in 2005.[6]

As mentioned earlier, McKay (2007 and this volume) has concluded that if we were to find, say, a microbial ecosystem on Mars, our ethical duty would be to "help" that system flourish and gain its fullest potential, in line with Leopold's and Naess's values placed on biodiversity. Similarly, for a lifeless planet McKay considers it the greatest good to terraform it so that it can support Earth life. Such actions clearly are not in accord with planetocentric ethics. My central concern is that we humans simply do not have (nor are likely to acquire at any time in the foreseeable future) the requisite knowledge and wisdom to "help" an ecosystem or planet gain its "fullest potential"—we are not even good at this kind of engineering on Earth, let alone with an exotic ecosystem. If we were to carry out such modifications, we would irreparably contaminate "nature's project" on Mars, so that it would be lost to scientific study forever. We gain more by observing nature's planetary experiments than by conducting our own, especially those that are irreversible. As Carl Sagan (1980) said:

> The [possible] existence of an independent biology on Mars is a treasure beyond assessing, and the preservation of that life must, I think, supersede any other possible use of Mars (109).
>
> If there is life on Mars, I believe we should do nothing with Mars. Mars then belongs to the Martians, even if they are only microbes (108).

Planetocentric ethics applies to all planetary bodies, but Rolston (1986) argues, and I agree, that special attention should be accorded (and regulations written for) certain bodies with properties that we find valuable for specific reasons. Examples might include Saturn and its rings and the Moon for their aesthetics; Comet Halley and the Moon for cultural reasons; Venus, Titan, and Io for their exotic surfaces; the Moon for its close relationship with Earth's history; Venus and Mars as Earth's companions that have traveled very different evolutionary paths over the past 4.6 Gyr; the asteroid belt and comets (all of them? or only selected samples?) as keys to understanding Solar System formation; and of course bodies deemed to be the best bets for harboring extraterrestrial life (Mars, Europa, Enceladus, and perhaps Titan). These examples are just first ideas, and there would need to be much negotiation as to categories and members of those categories.

Is Planetocentric Ethics Feasible?

Planetocentric ethics will not be an easy ethical scheme to put into force. It will require that society forgo past attitudes and radically rethink how it looks upon the valuation and exploration of new territory. Some may say that we can never change our values so fundamentally, and that economic and military forces will inevitably lead to the exploitation of planetary bodies once we have the technology to do so. Europa will be no different from Alaska. But long-held attitudes *can* change—witness how cigarette smoking in the United States over the past three decades has gone from a common activity with no restrictions to a relatively rare activity in public places. Or more importantly, witness how the attitudes of whites toward persons of color have fundamentally changed in the United States over the past half-century.

One good model for a planetocentric Solar System pact is the successful Antarctic Treaty of 1961, in particular its Protocol on Environmental Protection, added in 1998. Here we find the following wording, agreed to by all of the world's powerful nations (italics are mine):

> The Parties commit themselves to the *comprehensive protection of the Antarctic environment* and dependent and associated ecosystems and hereby designate Antarctica as *a natural reserve, devoted to peace and science.*
>
> Activities shall be planned and conducted in the Antarctic Treaty area so as to accord *priority to scientific research* and to *preserve the value of Antarctica* as an area for the conduct of such research, including research essential to understanding the global environment.

> The *protection of the Antarctic environment* and dependent and associated ecosystems and the *intrinsic value of Antarctica*, including its *wilderness and aesthetic values* and its *value as an area for the conduct of scientific research*, in particular research essential to understanding the global environment, shall be fundamental considerations in the planning and conduct of all activities in the Antarctic Treaty area.
>
> Activities in the Antarctic Treaty area shall be planned and conducted so as to avoid:
> (i) adverse effects on climate or weather patterns;
> (ii) significant adverse effects on air or water quality;
> (iii) significant changes in the atmospheric, terrestrial (including aquatic), glacial or marine environments;
> (iv) detrimental changes in the distribution, abundance or productivity of species of populations of species of fauna and flora;
> (v) further jeopardy to endangered or threatened species or populations of such species;
> (vi) degradation of, or substantial risk to, areas of biological, scientific, historic, aesthetic or wilderness significance.

Here is one possible transformation of this text for a Solar System Treaty (where "Solar System" refers more exactly to non-Earth locations) based on planetocentric ethics:

> The Parties commit themselves to the comprehensive protection of the Solar System environment and associated ecosystems and hereby designate the Solar System as a natural reserve, devoted to peace and science.
>
> Activities shall be planned and conducted in the Solar System Treaty area so as to accord priority to scientific research and to preserve the value of the Solar System as an area for the conduct of such research.
>
> Fundamental considerations in the planning and conduct of all activities in the Solar System shall be the protection of planetary bodies (planets, moons, asteroids, and comets) and associated ecosystems and the intrinsic value of the Solar System, including its wilderness and aesthetic values and its value as an area for the conduct of scientific research, in particular research essential to understanding the origin and history of the Solar System and its ecosystems.
>
> Activities in the Solar System shall be planned and conducted so as to avoid:

(i) adverse effects on the atmospheres, surfaces, or interiors of all planetary bodies;
(ii) changes in the distribution, abundance, or productivity of any indigenous species of organisms present; and
(iii) degradation of, or substantial risk to, areas of biological, scientific, historic, or aesthetic significance.

The Solar System is our last great wilderness. We must begin *now* to act to keep it untrammeled by human action, for we are on the cusp of producing major physical interactions with planets. We are like the New World colonists of the seventeenth and eighteenth centuries, who took on a vast wilderness ripe for exploitation. Today, however, we know where their actions ultimately led and are ethically bound to do much better.

Forty years ago the views of Earth brought back by the Apollo astronauts were groundbreaking in changing environmental attitudes on Earth. Now we can reverse directions as the resultant Earthly ecological ethics discussed in this chapter "rebound" outward and envelop the entire Solar System and its possibilities for extraterrestrial life.

Astrobiology and Solar System exploration were hardly in the mind of a nineteenth-century naturalist when he penned the following words, but they nevertheless work remarkably well even far from planet Earth:

> In Wildness is the preservation of the World.
>
> (Henry David Thoreau, "Walking," 1862)

Notes

1. Because we suppose that extraterrestrial *intelligence*, or in fact any form of life but the simplest, is absent from our own planetary neighborhood, my focus is on microbial life (or at most, simple multicellular organisms).
2. When I refer to *planets* or *planetary bodies*, I am including planets, satellites of planets, dwarf planets, asteroids, comets, and so on.
3. We should all join Chris's organization PETA: People for the Ethical Treatment of Aliens. Dues are very reasonable.
4. Note that even if something is determined to be a moral subject, this does not imply that it has rights, only that it has some degree of value to humans and should accordingly be rightly treated.
5. In fiction this is also related to *Star Trek*'s so-called Prime Directive, which prohibits interference with "pre-warp" civilizations.
6. The *Deep Impact* experiment is a good example of an activity that would undoubtedly *not* be acceptable under the proposed planetocentric ethics.

References

Callicott, J. B. 1986. "Moral Considerability and Extraterrestrial Life." In *Beyond Spaceship Earth: Environmental Ethics and the Solar System*, edited by E. C. Hargrove. San Francisco: Sierra Club Books.

Des Marais, D. J., J. A. Nuth, et al. 2008. "The NASA Astrobiology Roadmap." *Astrobiology* 8: 715–730.

Hargrove, E. C., ed. 1986. *Beyond Spaceship Earth: Environmental Ethics and the Solar System*. San Francisco, CA: Sierra Club Books.

Leopold, A. 1949. *A Sand County Almanac*. New York: Oxford University Press.

Lorenz, R. D., J. I. Lunine, and C. P. McKay. 1997. "Titan under a Red Giant Sun: A New Kind of 'Habitable' Moon." *Geophys. Res. Letters* 24: 2905–2908.

McKay, C. P. 1990. "Does Mars Have Rights? An Approach to the Environmental Ethics of Planetary Engineering." In *Moral Expertise: Studies in Practical and Professional Ethics*, edited by C. D. MacNiven. New York: Routledge.

———. 2007. "Planetary Ecosynthesis on Mars: Restoration Ecology and Environmental Ethics." Preprint (13 pages and 2 figures). Accessed June 21, 2012. http://esseacourses.strategies.org/EcosynthesisMcKay2008ReviewAAAS.pdf.

Naess, A., and G. Sessions. 1995. "Platform Principles of the Deep Ecology Movement." In *The Deep Ecology Movement: An Introductory Anthology*, edited by A. Drengson and Y. Inoue. Berkeley, CA: North Atlantic Books.

Race, M. S. 2007. "Societal and Ethical Concerns [in astrobiology]." In *Planets and Life: The Emerging Science of Astrobiology*, edited by W. T. Sullivan and J. A. Baross. Cambridge: Cambridge University Press.

Race, M. S., and R. O. Randolph. 2002. "The Need for Operating Guidelines and a Decision Making Framework Applicable to the Discovery of Non-Intelligent Extraterrestrial Life." *Adv. Space Res.* 30: 1583–1591.

Rolston, H. 1986. "The Preservation of Natural Value in the Solar System." In *Beyond Spaceship Earth: Environmental Ethics and the Solar System*, edited by E. C. Hargrove. San Francisco: Sierra Club Books.

———. 1988. *Environmental Ethics: Duties to and Values in the Natural World*. Philadelphia: Temple University Press.

Rummel, J. D. 2007. "Planetary Protection: Microbial Tourism and Sample Return." In *Planets and Life: The Emerging Science of Astrobiology*, edited by W. T. Sullivan and J. A. Baross. Cambridge: Cambridge University Press.

Sagan, C. 1980. *Cosmos*. New York: Random House.

Sullivan, W. T., and J. A. Baross, eds. 2007. *Planets and Life: The Emerging Science of Astrobiology*. Cambridge: Cambridge University Press.

United Nations Office for Outer Space Affiliates. 1979. "Agreement Governing the Activities of States on the Moon and Other Celestial Bodies." Accessed February 5, 2012. www.unoosa.org/oosa/SpaceLaw/moon.html.

U.S. Department of State. 1967. "Treaty on the Principles Governing the Activities of States in the Exploration and Use of Outer Space Including the Moon and Other Celestial Bodies." Accessed February 5, 2012. www.state.gov/www/globals/arms/treaties/space1.html.

CHAPTER TEN

Contact

Who Will Speak for Earth and Should They?

Jill Cornell Tarter
CENTER FOR SETI RESEARCH

For millennia humans have looked at the sky and wondered whether or not we are alone in this vast Universe. The modern tools of astronomy now give us the opportunity to try to find the answer to this old question; SETI, the search for extraterrestrial intelligence, is a scientific exploration that might provide an answer in the near future, much farther in the future, or never. But if one day this search (or some other methodology) detects evidence of a distant technological civilization, humans will have to decide whether and how we might communicate, and who will speak for Earth.

As we think about contact with other intelligent life, we do well to consider the physical facts first: where Earth lies in the Galaxy, and what this means in terms of connecting to any other life that might exist. Our home lies far out in the outskirts of the Milky Way, most of the hundreds of billions of other stars lie closer to the Galactic Center and far from us, and the tens of billions of other galaxies are much farther still. Connection and communication will be a long-distance affair. When and if we do eventually connect, what rules should we follow to converse? Who should decide the rules? What are the concerns for the larger society? And what should be scientists' concerns for this world's and the external world's society in such communication? Also, see Race's article in this volume for more general policy considerations when life elsewhere is finally detected.

Searching for Contact

Who cares about the search for extraterrestrial intelligence? Scientists are justifiably excited about searching for, and finding, life on Mars and other Solar System objects. But this life is expected to be microbial at best. Laypeople do not necessarily care about microbes but (as Hollywood proves repeatedly) they do care about intelligent life and contact with such life. Before considering the details of such contact, we need to think about how we even begin to recognize intelligent life elsewhere.

Basically we are looking for technology. If technology is short-lived—if a technological civilization develops and then dies out quickly—we will not find them, although we may detect their relic messages. If their technology is long-lived, we may find them. So the success of SETI depends on civilizations exhibiting long-lived technology. Consider these three facts, which highlight the great Galactic asymmetry (see figure 10.1):

- We are a very young technology in a very old Galaxy: 100 years versus 10 billion years.
- If a detectable technology exists elsewhere in the Galaxy, it is very likely to be older than we are.
- If the technology were younger, it would probably not be detectable (by us at this time).

Physicist Phil Morrison referred to SETI as the "archeology of the future" because a detected signal tells us about *their* past and the potential for *our* future. The detection of even a cosmic dial tone will tell us that it is possible to survive our technological adolescence.

But how do we search for their technology? In the theory of the structure of the human brain and functions of the mind, left-brain thinking is logical, sequential, objective, and analytical; our right-brain thinking is intuitive, holistic, subjective, and synthesizing. Our left brain tells us to search for analogs of terrestrial technologies—Earth on steroids; null results from such a systematic and extensive search could eventually be significant. By modeling what we are searching for, it is possible to delimit the volume of search space in which we did not find it, and to draw conclusions about whether or not that is significant. In searching the oceans for fish, the null results of scooping a single glass of water could not be considered significant. Our right brain tells us to search for markers of advanced technologies, those we ourselves have not yet conceived or invented. Our best chance of success from this right-brain approach is

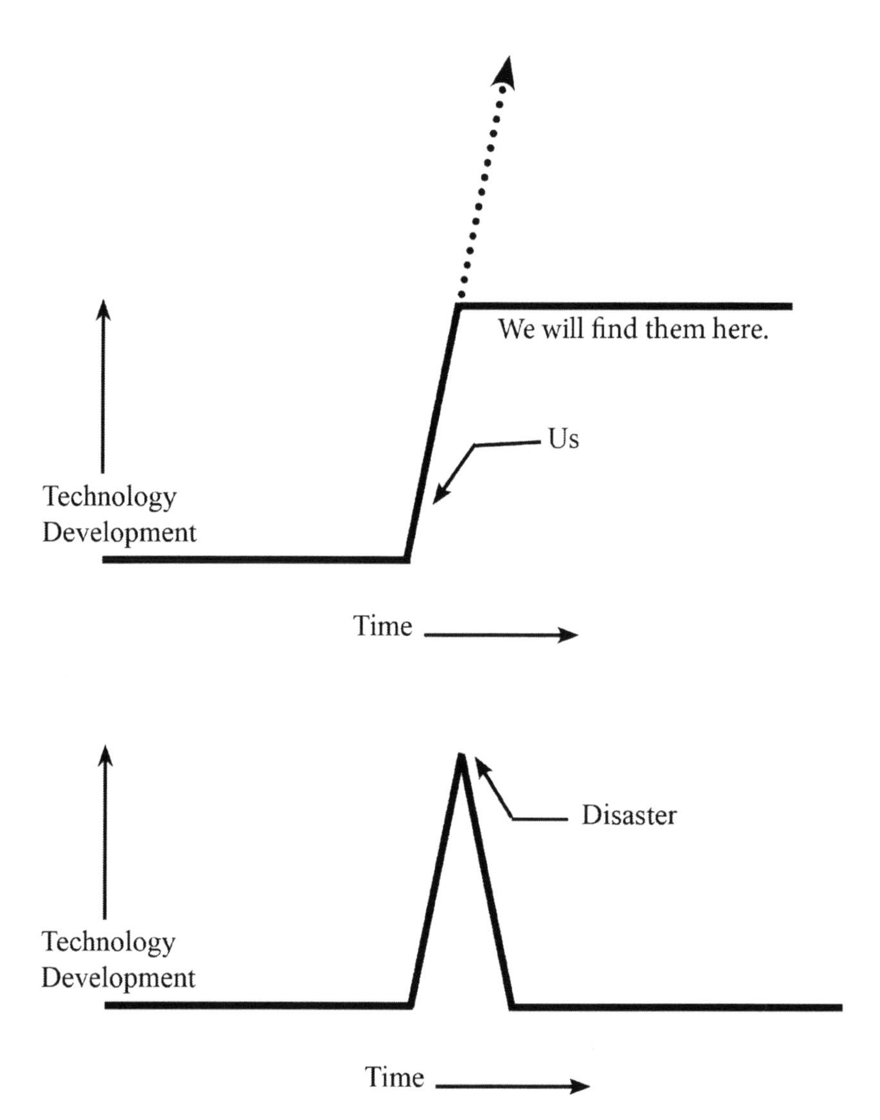

10.1 Any detectable technology will probably be older than we are. A long-lived technological civilization (upper image) will last over time, and we will find them. If a technological civilization is short-lived, we will not find them, as they will disappear before we find them.

to keep our astronomical eyes open for the unusual, the unexplainable. What advanced technologies could be out there? In this volume, Peters further considers the ways we might interact with an alien life form that we consider "the other."

The Unimaginable (Right Brain)

Arthur C. Clark's third law states: "Any sufficiently advanced technology is indistinguishable from magic" (Clarke 1973). Does this imply that we simply must give up, because we cannot design a search for magic? I would say no: Do not try to predict magic—just do astronomy! In just doing astronomy, wondrous and unexpected or unimagined discoveries do occur. For example, Martin Harwit's 1981 book, *Cosmic Discovery: The Search, Scope and Heritage of Astronomy*, documents the grand astronomical tradition of building a new instrument to open a new cell of observational phase space—and finding something totally unexpected. In 1967 Hewish and Bell discovered pulsars while working on identifying quasars and named the first one LGM-1 (Little Green Men-1) because the regular oscillations of its signal suggested it might have an intelligent origin. It is important to encourage today's graduate students and postdocs to pursue any "little bits of scruff" they may find in their data, and like Jocelyn Bell, they may make their own cosmic discovery.

The Imaginable (Left Brain)

Humans can imagine—and have imagined—some pretty interesting options for worlds harboring intelligent life. The idea of a *Machine World* arises from the fact that our digital technology is evolving exponentially and our carbon-based brain, evolving at the pace of biology, may soon be outperformed. Nanotechnology and machines may soon augment humans through some as yet unidentified process. Futurists call this the *singularity*. Hollywood calls this the *Borg*.

Another future is *Computer World*. In this scenario, information processing becomes the most important activity. Beings could deconstruct everything and then rebuild it into layers of computational capability. The beings' star could be surrounded in order to capture all energy, and the waste heat that comes from an inner layer would power the next layer out—matrioshka brains (a hypothetical structure of immense computing ability) of information processing.

What are the observable consequences of a system such as Computer World? First one would expect to observe infrared (IR) excess around such stars. Although astronomers do find many stars with IR excess, thus far we can explain this radiation as coming from planetary systems, from debris disks and proto-planetary disks around the stars; no matrioshka brains are required, at least not yet.

Perhaps in order to get rid of all the waste heat generated in an advanced Computer World and to drive up the efficiency of the computation, advanced technologies might place their computers (and themselves) in the coldest places in the Galaxy, near the edges, where even the weak interstellar starlight radiation field is further diluted. Such placement would produce strange, unexpected emission in these low-temperature regimes. It has been suggested that we focus our searching on the hinterlands of the Galaxy to find any such computing worlds.

A third alternative is *Power World*—in three variations. In 1964 Nikolai Khardashev developed a scale to classify civilizations based on their power production. A civilization of Type I manipulates the power incident on its planetary surface $\sim 4 \times 10^{19}$ ergs/s. Type II manipulates the total power output from its star $\sim 4 \times 10^{33}$ ergs/s. Type III manipulates all energy coming out of its galaxy $\sim 4 \times 10^{44}$ ergs/s. A Type I society might have space colonies in orbit to harvest more of the star's radiation. To capture all the star's energy, Type II civilizations might create a Dyson sphere, a system of orbiting solar satellites intercepting all the star's rays. A Type III society might harness the power of an entire galaxy: Kardashev envisioned them being able to warp space and time and thus able to exit from Euclidean space, or travel through it by means of a transportation system such as Carl Sagan envisioned in his novel *Contact*. Kardashev expected that a Type III civilization might thus be undetectable. It is hard to model what such a civilization might look like, though it has been suggested by science fiction writers that explosive Seyfert galaxies might be the industrial accidents of the cosmos.

More realistic observable consequences of Power Worlds include infrared (IR) excess around stars and coordinated nova or supernova explosions at large distances from each other. If a lot of fission reactions are used to produce power, advanced technologies might decide to safely dispose of fissile waste material by dumping it into their star—perhaps yielding enhanced emission lines of rare elements such as praseodymium or niobium in the spectrum of the star (if sufficient material is dumped). At one time, Przybylski's Star (HD 101065), one of the most peculiar stars known (its spectrum exhibited baffling emission lines), was thought to be a candidate for such a star, but theorists now prefer a more prosaic explanation involving line blanketing (a decrease in the star's spectrum due to unresolved absorption lines). Nuclear fusion plants produce neutrons and hence tritium, which is toxic (at least for humans), therefore an advanced technology might put such plants in orbit, thereby providing an opportunity to detect emission from the tritium by-product. Tritium has a half-life of 12.5 years, so if its line emission at 1516 MHz is detected any place except in the vicinity of a recent supernova, that radio signature could be the evidence of Power World technology.

Yet another imaginable world can be termed *Traveling World*. Advanced civilizations may seek to explore far from their home base. Books and movies postulate various modes of transportation—from the massive, galaxy-wide wormhole network of Sagan's *Contact* to the faster-than-light spaceships in much of science fiction. Nanoships and smart probes could be visiting our Solar System at the present time.

Transportation systems provide a variety of observational consequences. In the movie *Contact*, 18 hours of recorded static provided evidence for the reality of an extended trip that appeared instantaneous to the observers, and in the book version the advanced wormhole-building civilization provided a message in digits of π. Gamma-ray bursts are enigmatic phenomena indicating massive explosions. Perhaps matter-antimatter annihilation rockets explain these tremendous energy bursts, and attempts have been made to search for four-dimensional space-time alignments indicative of trajectories. Asteroids might be sources of fuel for long-slow interstellar ships or be ships themselves. Papagiannis (1978) suggested that double asteroids might be evidence of such spacefaring technologies. Ida and Dactyl are one such observed double, but radar imaging confirms the more mundane nature of these rocks. Finally, nanoprobes might be lurking at Lagrangian points where reflected sunlight could make them detectable. The sensitivity of the observations of these gravitational minima to date are adequate to rule out the presence of big, bright, reflective objects, but not small, artificial objects.

The final approach one might imagine for detection of a distant technological civilization is *Beacon World*. Signals deliberately transmitted from such a world might look "almost natural" and be captured during routine astronomical explorations, or conversely they might appear obviously engineered and clearly not due to astrophysics. In the almost-natural category might be a "pulsar" that upon further analysis was recognized to regularly switch its period between two values, unlike a natural pulsar. A blinking star might indicate artificial transits created by orbiting very large screens in shapes different from a circular planetary cross section. The light curves from the *Kepler* spacecraft and other planet-searching missions contain enough information during limb crossing to discover any such artificial transits. Perhaps the regular period of a Cepheid variable star could be technologically advanced, so that it would appear early in its phase curve. This normal and advanced phase variation could then serve as an omnidirectional beacon for a Morse code-like interstellar message.

Such almost-natural signals might be serendipitously discovered by any young civilization beginning to explore its astronomical surroundings. Deliberate beacons could also be transmitted at single radio frequencies. This non-natural, frequency compression model has been the basis for the majority of SETI searches since 1960. More recently SETI researchers have also begun to look for signals exhibiting extreme time compression, targeting nearby stars and seeking very bright, very short optical laser pulses that last for a nanosecond or less. The proof of this concept is the fact that we already have high-powered lasers that would rival the Sun's brightness if directed and seen from afar.

The observational consequences for Beacon Worlds are the construction of dedicated telescopes and detectors for the detection of obviously engineered beacons. For almost-natural signals, just doing astronomy could reveal their true nature in time.

The Search So Far

How much of the sky have we explored? To date astronomers have discovered over 900 planets around other stars—almost all within 300 light-years of the Sun (see chapter 1 for more details on the successful search for exoplanets). The SETI Institute's Project Phoenix spent a decade looking for signals from about a thousand stars out to 200 light-years. Optical SETI projects at Harvard, Berkeley, and Lick Observatory have probed several thousand nearby stars. The Allen Telescope Array (ATA), a radio interferometer now starting to observe in northern California, will spend a decade looking at a million stars out to 1000 light-years—still a very small part of the Galaxy. The ATA will also survey 20 square degrees surrounding the Galactic Center, thereby exploring many billions of very distant stars for the very powerful type of transmitters envisioned by Kardashev. SERENDIP V (Search for Extraterrestrial Radio Emissions from Nearby Developed Intelligent Populations spectrometer installed at Arecibo) is beginning to use the seven-element focal plane array at Arecibo to continue a random radio sky survey of the 30 percent of the sky it can access, and involve the public in the SETI@home distributed computational analysis of its data.[1] The Harvard OSETI project is now surveying 80 percent of the sky visible from Massachusetts.

If ATA and these other programs plus their future elaborations find no signals in the next decade, or the decade after that, do we give up? Only if we have not invented new technologies to try, and only if we have made such substantial improvement over today's single glass of water sample of the cosmic ocean, that the null results are significant. SETI will succeed, and then transition from the discovery phase to the stamp-collecting phase as we try to better understand our place in the cosmos, or it will end when the public is no longer willing to bear even the relatively small amount of funding invested in the searches, when the effort that has been expended is commensurate with the significance of the conclusion of cosmic isolation.

Assuming we will detect signals, we need to consider how we view those technologists, or their surrogates that send the signals. If we do make contact with civilizations that have created any of these technologies, what possible reactions may occur? Our left brain says: *They are "like us."* This assumption means that we will expect some understanding and attribute to them a status of moral agent and subject. George Wald at the conference Life Beyond Earth and the Mind of Man, in 1972, said, "I can conceive of no nightmare as terrifying as establishing such communication with a so-called superior (or if you wish, advanced) technology in outer space." Our left brain says we will err because the aliens appear similar to us so we proceed based on false assumptions. Think of how humans react to and communicate with the familiar dogs and the unfamiliar cuttlefish.

Our right brain says: *They are unimaginably different and strange.* We are forced to acknowledge their longevity. Their ethical status is not intuitive. Their very strangeness may help us act "appropriately"—if we can avoid a belligerent response to the "other," which would be irrational in view of their longevity.

A Rational and Ethical Approach to Contact—A Set of Principles

In 1989 the International Academy of Astronautics (IAA) produced the *Declaration of Principles concerning Activities Following the Detection of Extraterrestrial Intelligence* (IAA 1989). The declaration has nine sections outlining procedures to follow if detection occurs. These principles dealt only with the detection of some sort of electromagnetic evidence; they did not deal with the eventuality of physical contact. Nevertheless, as shorthand, conditioned by historical science fiction, we will use the term "contact" for the detection of electromagnetic evidence. Such contact would be unprecedented in the history of humanity, with only superficial similarity to contact with persons and species between Earth's continents. The declaration seeks to establish ethical conduct in reporting such a discovery to humanity, but scientists and lawyers created this declaration; it deals with conduct involved in sharing information, not the profound ethical and theological issues that will arise from contact. The nine steps to follow in case of detection are these:

1. If contact seems to have occurred, seek to verify.
2. If verified locally by the discovery team, make contact with parties to this declaration to obtain additional help to verify.
3. If contact seems credible, inform observers worldwide through the Central Bureau for Astronomical Telegrams of International Astronomical Union. [And in the twenty-first century, discretely notify those funding the effort.]
4. Make a formal public announcement of the discovery with care being given to attribution of all involved.
5. Publish the discovery data.
6. Continue to monitor and record data from the source.
7. Ask the International Telecommunication Union for extraordinary protection of the signal's frequencies (if the evidence lies in the radio regime).
8. No response to the signal should be made without global consensus outlined in a separate declaration.
9. A task force should try to keep control of the process.

How to Make and Verify Contact

The SETI Institute's first SETI project (Phoenix) was the most sensitive and sophisticated ever, but it had limited telescope time (spread over nine years) resulting in total observing of only 482 days (15 percent of total observing time), it had limited frequency coverage (1.2–3.0 GHz), it looked at only about 800 stars to a distance of about 200 light-years, and it detected no signals. The project used three facilities: the Parkes 210-foot radio telescope in New South Wales, Australia; the National Radio Observatory's 140-foot telescope in Green Bank, Virginia; and the largest radio telescope in the world, in Arecibo, Puerto Rico. To verify whether a signal was extraterrestrial or earthly interference, this project used a second telescope located hundreds of miles from the primary detection instrument. Because the signal type being sought was narrowband, differential Doppler shifts relative to the distant source would cause offsets in both frequency and frequency drift as measured by these two widely separated telescopes. The differences could be precisely calculated, assuming the signals were originating from the target star. The wrong rates of drift indicate a terrestrial overflying or orbital technology, and no frequency drift indicates a signal local to the observatory.

The Allen Telescope Array is composed of many telescopes. The first of four construction phases of the ATA is complete, with an array of 42 antennae of the total 350. The spacing of these telescopes ranges from 10 meters to 900 meters. ATA targets about 250,000 stars selected from HabCat, a Catalog of Nearby Habitable Systems (Turnbull and Tarter 2003a, 2003b), created from what is known about stars in the Hipparcos and Tycho catalogs, and selecting "habstars," stars that are likely to be good hosts for habitable planets, near the Sun.[2]

The search on ATA employs the next-generation signal detector SonATA (SETI on the ATA), which uses three beams (virtual telescopes) within the large field of view of the array and observes three stars at the same frequency at the same time. A signal that appears to come from two or three stars is interference entering through the array side lobes; a signal that appears to come from only one star is interesting. As the signals from the 42 telescopes are added together with the correct time delays and phase shifts to pick out a particular beam within the field of view, slightly different weightings can also create a simultaneous null somewhere else on the sky. To check that a signal detected in two beams is in fact not due to a strong SETI signal from one of the two stars picked up at lower sensitivity in the beam of the other star, an offset null formed along with the beam on one star is placed on the position of the other star.

The ATA began commensal targeted SETI searches (foreground target stars are observed with beam-formers at the same time that radio astronomical maps of the entire field of view are generated with a spectral imaging correlator) and a survey of the Galactic Center region in the spring of 2009.[3] In February 2009, Jill Tarter was awarded a TED Prize, including the opportunity to make a wish to change the world. That wish is now being shaped into OpenSETI, with a number of components that will allow open source developers, digital signal processing experts, and everyone with a willing set of eyes to participate in making the SETI searches on the ATA even better. If successful, OpenSETI will form powerful social networks that can inform and involve the global community, ultimately changing its perspective to a more appropriate cosmic, rather than national, point of view: Earthlings all.[4]

The Harvard OSETI targeted search has made over 16,000 observations for nanosecond-pulsed optical signals since the start of its operations in 1998, in conjunction with an optical program to search for extrasolar planets. Today the Harvard program uses an innovative optical telescope, designed and built by Paul Horowitz and his students. It surveys the northern sky as it drifts overhead utilizing a 16 × 64 photodiode array, which provides a sensitivity of 20 photons/m^2/ns, a fivefold improvement over the targeted search. To follow up on some interesting detections, the Whipple 10-m telescope observed 12 areas of sky for one to seven hours per area using an algorithm to extract three-pixel "single pixel" events with an improved sensitivity of 10 photons/m^2/50 ms with no repeated events.[5] Imaging Cherenkov telescopes such as the Whipple Observatory began undertaking SETI projects in the early 2000s (Holder et al. 2005).

The SERENDIP observations at the Arecibo Observatory (currently SERENDIP V) take advantage of the unusual optics of the antenna, conducting a random SETI survey of the sky overhead while astronomers are doing their own surveys. A simple event-above-threshold detection takes place in real time at the observatory. In addition, SETI@home records the data, and a more sensitive set of detection algorithms (pattern recognition in frequency and time, similar to those done with the ATA) is executed on this recorded data offline with the help of a global team of volunteers who run a background screensaver on their machines.[6]

The observatories above and several others around the world are involved in SETI projects, and can verify the original observation. Those "on the air" include ATA, SETI Italia, SETI League's Project Argus, Harvard OSETI Sky Survey, Southern SERENDIP Argentina, Lick Observatory, and Leuschner Observatory. To verify the signal, the observatory needs

- a comparable radio or optical telescope—size matters, but integration time can help in radio wavelengths;
- receivers to cover the frequency;
- back ends sensitive to signal type; and
- a location west of the discovery site.

Radio astronomy observatories may be able to use "garden variety," commercial spectrum analyzers, but fast photon counters are likely to be accessible only at another OSETI site.

Going Public with the News

Before going public, the discovery team should prepare the press kit using the KISS (Keep It Simple, Stupid) criterion. Copying from the Richter

The Rio Scale: RS = Q* δ

Q (Level of Confidence = class + type + apparent distance)			
Class	Type	Distance	
Traces of astroengineering	Archival data, aposteriori	Extragalactic	1
Leakage, non-decipherable	From other observations, transient	Within the Galaxy	2
Omni beacon, no information	SETI/SETA, verified, transient	Close enough for communication	3
Earth-specific beascon, no info	From other observations, repeatable	Within Solar System	4
Earth-specific information or encounters			

δ (Credibility of Report)	
Obvious hoax or fraudulent	0
Very uncertain, try to verify	1 out of 6
Possible, but requires verification prior to acceptance	2 out of 6
Probable, has already been verified	3 out of 6
Absolutely reliable, without doubt	4 out of 6

Rio Scale	
Extraordinary	10
Outstanding	9
Far-reaching	8
High	7
Noteworthy	6
Intermediate	5
Moderate	4
Minor	3
Low	2
Insignificant	1
None	0

10.2 The Rio Scale: RS = Q*δ. SETI researchers devised a numerical scale for categorizing reported signal detections. Weighted values combine to produce a scale of 0–10 to indicate the importance of the detection.

scale, SETI researchers have devised a system for categorizing reported signal detections, the Rio Scale (RS) = Q*δ, where Q = level of consequence = class + type + appearance, and δ = credibility of report. This formula creates a simple numerical indicator of the importance of the claimed detection (see figure 10.2). Although this metric is virtually unknown at the moment, it is the intention of the SETI community to seek opportunities for publicizing it—for example, by using it to "score" new novels and movies dealing with a contact scenario. Keeping it simple, and in a familiar format, will lead to better understanding and, one hopes, less potential for exploitation in encountering ETI.

Scientists do not always consider the effects of their discoveries, but contact with extraterrestrial intelligence will have a profound effect on humanity. It will change forever the way we think of ourselves. Making the discovery data themselves, and the observations, accessible and making the data reduction understandable are critical tasks—they may not ensure that understanding is achieved, but without care in these critical tasks and recognition of the import of this discovery and how to communicate it, understanding and acceptance will be much more difficult. In the absence of a local source of authority to serve as an interpreter of the actual discovery and its probable consequences, the media will tend to construct their own fiction. That is why the SETI investigators suggest the use of the International Astronomical Union (IAU) Telegram System. Nearly every observatory in the world subscribes to this time-critical alert system. By getting the information into the hands of observers close by, SETI investigators alert local resources to prepare for inquiries This action provides knowledgeable sources for journalists anywhere in the world.

In terms of a unified global response to contact, we return to the Declaration of Principles. Section VIII states:

> No response to a signal or other evidence of extraterrestrial intelligence should be sent until appropriate international consultations have taken place. The procedures for such consultations will be the subject of a separate agreement, declaration or arrangement.

Given the current political conditions of a world full of disparate states, philosophies, and individual agendas, there is unlikely to be any global consensus. The principles have no power of enforcement, and it is highly likely that this section of the Declaration will be ignored, for a number of reasons:

- No global governance exists;
- He who pays the transmitter gets to transmit;
- the International Telecommunication Union controls the spectrum on the basis of international treaty, but local political entities are charged with enforcement;
- Sullivan, Brown, and Wetherill (1978) demonstrated that much information about the physical environment of the transmitter can be garnered indirectly from unintentional leakage radiation.

Freeman Dyson (1997) argues that the conditions of our world in the twenty-first century, and the nature of humans, would best be demonstrated by the very character of the uncontrolled cacophony from all responders. So if we want to send a message displaying the true condition of our world, perhaps multiple, uncoordinated, and uncontrolled responses should be encouraged. But there are those who worry about *any* response.

Keeping Control of the Process

If contact comes as a result of either listening or active communication, how likely is it that any task force can keep control of the process? Item number 9 in the Declaration of Principles seems the least likely to be achievable. Despite the likely outcome of lack of control, there has been some subsequent thought about how to use existing space law to guide the postdetection process.

Two previously executed international treaties considered the possibility of the detection of microbial life, but their sections could possibly be extended to apply to responses to ETI:

- Article V (3) of the 1967 Outer Space Treaty
  (information of phenomena which could constitute a danger)
- Article 6 (3) of the 1979/1984 Moon Agreement
  (information of indication of organic life)

Further deliberations of the IAA and the International Institute of Space Law (IISL) subsequent to the 1989 Declaration of Principles produced several additional outcomes and recommendations concerning detection:

- Special issue of *Acta Astronautica* in 1990 (Tarter and Michaud 1990);
- IAA Position Paper on Transmissions to Extraterrestrial Intelligence (ETI) (1998) approved by the IISL: End note 3 provides a new draft Declaration of Principles[7] that states:
  - The United Nations (UN) and its Committee on the Peaceful Uses of Outer Space (COPUOS) is the most suitable body to take the lead;
  - All interested states could participate;
  - Any message should be sent on behalf of all humankind;
  - Any message should broadly represent humanity; and

- Scientists, scholars, engineers, and other specialists could serve as consultants.

The UNCOPUOS has been formally briefed on the 1998 IAA Position Paper and has outlined and recorded its concerns about and its role in SETI:

- The detection of ETI will have a profound impact on human civilization;
- Improved technologies are being used and interest in SETI is increasing worldwide;
- The detection could happen at any time—or not for a very long time; and
- As the most authoritative international space body, the UNCOPUOS, might be called upon to act in response to a detection.

Deliberations by the United Nations tend to be glacially slow. Action is initiated by the General Assembly, and to date no heads of Nation State members have responded to requests to place this issue on that body's agenda.

Should We Broadcast?

It is one thing to receive transmissions *from* "out there," but it is another thing altogether to send transmissions *to* "out there." Receipt is essentially passive (although it requires active listening), but broadcast is willfully advertising humanity's presence to the cosmos. In the early 1990s, SETI researchers held a series of workshops about the societal implications of contact, including the issue of broadcasting humanity's presence to the cosmos. (This follows the standard scientific dictum, which is often wise: When in Doubt—Hold a Workshop.) At the workshops, representatives of sciences (including social sciences), history, media, several major religions, and policy met to discuss the issue. These "representatives" were, however, all white, mostly male, and inhabitants of the industrialized world only. The attendees produced the report, "Social Aspects of the Detection of Extraterrestrial Civilization—A Report of the Workshops on the Cultural Aspects of SETI held in October 1991, May 1992, and September 1992 at Santa Cruz, California."

Some of the workshop participants were also involved in drafting the Declaration of Principles. Thus the prevailing point of view was that control and prohibition of transmissions were impractical and impossible, so preparation and advance planning were the most prudent policies. The workshop participants were acutely aware of their limited representation of the cultures, belief systems, and knowledge systems of the globe, and it was their recommendation that the discussions be extended to a broader, global audience. Although the logistical and financial support has never been found within the SETI community for such a face-to-face global meeting, new technologies and viral social networking applications may enable a global discourse in ways never imagined by the first workshops.

In yet another series of workshops to develop a technical roadmap for SETI at the SETI Institute for two decades to come (published as *SETI 2020* in 2001), the participating scientists, engineers, and technologists had a somewhat different perspective on the question of *ab initio* transmissions or broadcast replies to a detected signal. They reflected that twenty-first-century humans have an asymmetric relationship with the Universe: we are a very young technology (~100 years) in a very old Galaxy (~10 billion years). Any technology that we can detect is going to be much older than we are, and we should follow their lead. That asymmetry led the workshop participants to these detailed conclusions:

- If or when we achieve contact with another civilization, that civilization will certainly be more technologically advanced than we are. Contact with a less technologically advanced civilization is not now a possibility. In fact, any civilizations we contact are statistically likely to be far more advanced. When the evolution of planets and their attendant technologies require billions of years, it is unlikely that two technological civilizations will be synchronized to better than a million years.
- If it happens at all, there always has to be a *first* contact between two technological civilizations. Statistically, it is extremely unlikely that our first contact with an ETI civilization will also be its first contact with an ETI civilization. Thus, the advanced technology we detect will have experienced this type of encounter many times before. It already may have established a galactic protocol for information interchange, to which *ab initio* transmissions by Earth will have no chance of adhering. Thus, we justify our asymmetrical *listen only* strategy by recognizing our asymmetrical position among galactic civilizations. We are one of the youngest!
- Transmitting is a more expensive strategy than receiving. Within the next two decades, the parameter space explored for signals can be extended by the compounded growth rate of many technologies.

Transmissions could benefit from these same exponential improvements in technology, but with the limited resources likely to be available during this same period, we could not add significantly to the high power of our leakage radiation. As that leakage abates or becomes more noiselike, this argument loses its force. Transmission will not be rewarded for decades, perhaps centuries, because of the great distances and round-trip travel times for signals. Our resources are constrained, and it is thus prudent to pursue a passive program of exploration that might provide a positive result within years.

- Transmission is a diplomatic act, an activity that should be undertaken on behalf of all humans. We lack the cultural maturity to accomplish such a cohesive action. Some Working Group participants felt strongly that this active strategy should not be embarked upon unilaterally, without consultation and consent. Although most of the participants believed that transmitting now would be merely harmless and wasteful, a few members felt that transmissions should not be carried out without international consultation and approval by appropriate international administrative bodies.

So once again those directly involved with SETI concluded that a global discussion was called for, although they suggested that an advanced technological civilization would ultimately set the agenda for interaction based on its previous experience.

These workshops more or less ignored the fact that we have already deliberately sent humanity's signals to the cosmos—without consensus even among scientists as to the merit of the acts. In 1974, as part of the dedication ceremony for a new, upgraded antenna surface, researchers sent a message from Arecibo Observatory to the globular cluster M13 at a distance of 25,000 light-years. This broadcast provoked a variety of responses, including a rebuke from Astronomer Royal Sir Martin Ryle: "Any creatures out there may be malevolent or hungry." Subsequent workshop participants disregarded this transmission because they understood that it was a stunt—its short time duration (<10 minutes) made it extremely unlikely that it would ever be detected. A recipient would have to be looking at the Earth with precisely the right technology at precisely the 10-minute interval during which the signal passed over them; our daily leakage of broadcast TV and radio programs makes us more detectable, for now. If a deliberate transmitting effort is to have any reasonable chance of establishing contact by making Earth detectable to another technological civilization, it will need to persist for a period of time that is long in a cosmic context. This is something we are not now capable of undertaking.

Four "messages in bottles" have also been sent aboard spacecraft. *Pioneer 10* and *Pioneer 11* held gold anodized plaques with pictorial messages showing female and male humans and information about where and when in the Galaxy these spacecraft had originated. *Voyager 1* and *Voyager 2* carry golden records with the sights and sounds of Earth encoded and were intended to provide samples of the diversity of life and culture on Earth—at least as they existed in 1977 and as chosen by the Voyager team. But space is so empty that these bottled messages will not approach close to any other star systems for more than 100,000 years. By that time there will be no onboard activity, as the thermonuclear power sources will have decayed, and these artifacts will be very difficult to discover among the (presumed) large number of small bodies comprising any planetary system.

Finally, enterprising individuals and even NASA can send their own messages. Sentforever.com advertises messages beamed into space from a radio transmitter for the bargain price of £9.95 (US$15.50) with tracking an option. On February 4, 2008, to celebrate the fiftieth anniversary of the founding of NASA and the 40th anniversary of the day the song was written, NASA used a Deep Space Network transmitter to send the Beatles song "Across the Universe" in the direction of the North Star, Polaris. Again, these messages are weak and transitory and do not noticeably augment our current detectability from leakage. Similar to the Rio Scale for assessing the potential significance of a reported signal detection, transmitted signals have their own scale, the San Marino Index (SMI) = I + C where I = intensity of transmission and C = character of transmission. With more scales than signals, we perhaps "signal" our special form of hubris.

Even with these previous broadcasts and workshop deliberations, the questions remain to be answered: Should we broadcast? Who should speak for Earth? What should they say? Serious efforts at transmission require long-term commitment and global cooperation and coordination—all of which are still lacking. As previously mentioned, new technologies of social networking may improve the possibilities. The "Long Now" Clock and Library offer the first real opportunity to demonstrate a long-term commitment to a project that has a timescale that is relevant to a deliberate transmission effort.[8] Although the Long Now Foundation has not taken up the question of incorporating a transmitter and has no current plans to do so, those with a long view representing 10,000 years of humanity's records may be the appropriate messengers:

- Who should speak for Earth? Those who wind the clock.
- What should they say (transmit)? The library of humanity.

Conclusion

Perhaps the best way to address SETI—the most ethical, the most useful, the most appropriate—is to acknowledge the asymmetry: humanity is in its infancy compared to any ETI encountered. At least for the near term, an active SETI program with message transmission does not need to be pursued. Transmission may be the strategy of choice after (and if) our technology grows up and can deliver on very long-term projects. For now, listening will suffice. If detection does occur as a result of listening efforts, humans should follow ETI's lead, for it is statistically certain they have done this before.

How humanity will face its fall from its self-proclaimed pedestal will be extraordinary to witness—maybe trauma from which there is no recovery, maybe euphoria and a new stage in human development. Or then again, maybe busy people will do nothing more than issue a collective yawn. Regardless of what results, contact will expand humanity's views of society and self unlike any other event in history. Scientists and others have started efforts to deal with this event in principled ways, in mechanisms that can only form a foundation for profound considerations not addressed in the realm of science.

Notes

1. SETI@home is a scientific experiment that uses Internet-connected computers in the search for extraterrestrial intelligence (SETI). Anyone can participate by running a free program that downloads and analyzes radio telescope data. Currently it is the largest distributed computing effort, with over three million users worldwide. For more information about SETI@home, see http://en.wikipedia.org/wiki/SETI@home.
2. The Hipparcos and Tycho catalogs are the primary products of the European Space Agency's astrometric mission, Hipparcos.
3. See www.seti.org/ATA.
4. See www.tedprize.org/jill-tarter/.
5. See http://seti.harvard.edu/oseti/.
6. See http://setiathome.ssl.berkeley.edu/.
7. *Declaration of Principles concerning Activities Following the Detection of Extraterrestrial Intelligence:*

    1. Any individual, public or private research institution, or governmental agency that believes it has detected a signal from or other evidence of extraterrestrial

intelligence (the discoverer) should seek to verify that the most plausible explanation for the evidence is the existence of extraterrestrial intelligence rather than some other natural phenomenon or anthropogenic phenomenon before making any public announcement. If the evidence cannot be confirmed as indicating the existence of extraterrestrial intelligence, the discoverer may disseminate the information as appropriate to the discovery of any unknown phenomenon.

2. Prior to making a public announcement that evidence of extraterrestrial intelligence has been detected, the discoverer should promptly inform all other observers or research organizations that are parties to this declaration, so that those other parties may seek to confirm the discovery by independent observations at other sites and so that a network can be established to enable continuous monitoring of the signal or phenomenon. Parties to this declaration should not make any public announcement of this information until it is determined whether this information is or is not credible evidence of the existence of extraterrestrial intelligence. The discoverer should inform his/her or its relevant national authorities.
3. After concluding that the discovery appears to be credible evidence of extraterrestrial intelligence, and after informing other parties to this declaration, the discoverer should inform observers throughout the world through the Central Bureau for Astronomical Telegrams of the International Astronomical Union, and should inform the Secretary General of the United Nations in accordance with Article XI of the Treaty on Principles Governing the Activities of States in the Exploration and Use of Outer Space, Including the Moon and Other Bodies. Because of their demonstrated interest in and expertise concerning the question of the existence of extraterrestrial intelligence, the discoverer should simultaneously inform the following international institutions of the discovery and should provide them with all pertinent data and recorded information concerning the evidence: the International Telecommunication Union, the Committee on Space Research of the International Council of Scientific Unions, the International Astronautical Federation, the International Academy of Astronautics, the International Institute of Space Law, Commission 51 of the International Astronomical Union, and Commission J of the International Radio Science Union.
4. A confirmed detection of extraterrestrial intelligence should be disseminated promptly, openly, and widely through scientific channels and public media, observing the procedures in this declaration. The discoverer should have the privilege of making the first public announcement.
5. All data necessary for confirmation of detection should be made available to the international scientific community through publications, meetings, conferences, and other appropriate means.
6. The discovery should be confirmed and monitored and any data bearing on the evidence of extraterrestrial intelligence should be recorded and stored permanently to the greatest extent feasible and practicable, in a form that will make it available for further analysis and interpretation. These recordings should be made available to the international institutions listed above and to members of the scientific community for further objective analysis and interpretation.

7. If the evidence of detection is in the form of electromagnetic signals, the parties to this declaration should seek international agreement to protect the appropriate frequencies by exercising procedures available through the International Telecommunication Union. Immediate notice should be sent to the Secretary General of the ITU in Geneva, who may include a request to minimize transmissions on the relevant frequencies in the Weekly Circular. The Secretariat, in conjunction with advice of the Union's Administrative Council, should explore the feasibility and utility of convening an Extraordinary Administrative Radio Conference to deal with the matter, subject to the opinions of the member administrations of the ITU.
8. No response to a signal or other evidence of extraterrestrial intelligence should be sent until appropriate international consultations have taken place. The procedures for such consultations will be the subject of a separate agreement, declaration or arrangement.
9. The SETI Committee of the International Academy of Astronautics, in coordination with Commission 51 of the International Astronomical Union, will conduct a continuing review of procedures for the detection of extraterrestrial intelligence and the subsequent handling of the data. Should credible evidence of extraterrestrial intelligence be discovered, an international committee of scientists and other experts should be established to serve as a focal point for continuing analysis of all observational evidence collected in the aftermath of the discovery, and also to provide advice on the release of information to the public. This committee should be constituted from representatives of each of the international institutions listed above and such other members as the committee may deem necessary. To facilitate the convocation of such a committee at some unknown time in the future, the SETI Committee of the International Academy of Astronautics should initiate and maintain a current list of willing representatives from each of the international institutions listed above, as well as other individuals with relevant skills, and should make that list continuously available through the Secretariat of the International Academy of Astronautics. The International Academy of Astronautics will act as the Depository for this declaration and will annually provide a current list of parties to all the parties to this declaration.

8. See www.longnow.org/.

References

Clarke, A. C. 1973. *Profiles of the Future: An Inquiry into the Limits of the Possible.* New York: Phoenix. Revised edition, 2000.

Dyson, F. J. 1997. *Imagined Worlds (Jerusalem-Harvard Lectures).* Harvard, MA: Harvard University Press.

Harwit, M. 1981. *Cosmic Discovery: The Search, Scope and Heritage of Astronomy.* Cambridge, MA: MIT Press.

Holder, J., P. Ashworth, S. LeBohe, H. J. Rose, and T. C. Weekes. 2005. "Optical SETI with Imaging Cherenkov Telescopes." *29th International Cosmic Ray Conference Pune* 5: 387–390.

International Academy of Astronautics. 1989. "Declaration of Principles concerning Activities Following the Detection of Extraterrestrial Intelligence." *Acta Astronautica* 21 (2): 153–154.

Kardashev, N. 1964. "Transmission of Information by Extraterrestrial Civilizations." *Soviet Astronomy* 8: 217.

Papagiannis, M. 1978. "Are We All Alone, or Could They Be in the Asteroid Belt?" *Quarterly Journal of the Royal Astronomical Society* 19: 277–281.

Sagan, C. *Contact*. 1985. New York: Pocket Books.

Tarter, J. C. 2001. "The Search for Extraterrestrial Intelligence (SETI)." *Annual Review of Astronomy and Astrophysics* 39: 511–548. doi: 10.1146/annurev.astro.39.1.511.

Tarter, J. C., and M. A. Michaud, eds. 1990. "SETI Post Detection Protocol." Special issue, *Acta Astronautica* 21 (2).

Turnbull, M. C., and J. C. Tarter. 2003a. "Target Selection for SETI. I: A Catalog of Nearby Habitable Stellar Systems." *Astrophysical Journal Supplement* 145: 181–198.

———. 2003b. "Target Selection for SETI. II: Tycho-2 Dwarfs, Old Open Clusters, and the Nearest 100 Stars." *Astrophysical Journal Supplement* 149: 423–436.

CHAPTER ELEVEN

Astroethics

Engaging Extraterrestrial Intelligent Life-Forms

Ted Peters
PACIFIC LUTHERAN THEOLOGICAL SEMINARY
AND THE GRADUATE THEOLOGICAL UNION

How might we frame ethical concerns that are likely to arise when we begin to engage interactively with extraterrestrial intelligence? Based upon terrestrial experience, can we export Earth ethics? To speak to such questions, we must speculate. Ethical deliberation in light of astrobiology will unavoidably extrapolate and imagine scenarios both in continuity and in discontinuity with what we are already familiar.

Is it realistic to imagine that life exists elsewhere in our cosmos, either within our Solar System, within the Milky Way, or beyond? "There is growing scientific confidence that the discovery of extraterrestrial life in some form is nearly inevitable," say Margaret Race and Richard Randolph (Race and Randolph 2002). "Almost beyond doubt, life exists elsewhere," writes David Darling (2001).

Might we at some point in the future find ourselves engaging with extraterrestrial intelligent life (ETIL)? Assuming the answer to be yes, I would like to speculate on appropriate frameworks for pursuing *astroethics*, what others might refer to as "space ethics" (Arnould 2005). The field of astroethics would encompass our search for extraterrestrial nonintelligent life (ETNL) forms, the terraforming of other habitats within our Solar System, listening for radio signals sent by intelligent civilizations, and imagining an even more speculative scenario in which all living creatures on Earth would find themselves working out a new relationship with an extraterrestrial civilization. It is the final item—engaging ETIL—that I will address in this chapter. In anticipation of contact with ETIL in a form that reasonably resembles *Homo sapiens* on Earth, perhaps we could consider engagement with three possibilities: extraterrestrial biotic individuals who are inferior to us (less evolved), our peers (equally evolved), and superior to us (more highly evolved). Each of these three categories implies different foundations upon which to construct an appropriate ethical superstructure.

The ethical orientation here will derive from a sense of moral responsibility. As suggested by the etymology of the Latin (*respondere* meaning "to answer"), responsibility ethics answers questions raised by our changing situation (Jonsen 1986). Establishing a new relationship with extraterrestrials would prompt many questions; and an ethic of responsibility would seek to spell out just how it would be best for earthlings to respond. Further, the idea of responsibility includes care, care for both the health and the welfare of planetary life on Earth but also the health and welfare of our new space neighbors. The conditions and imperatives arising from the new situation will suggest forms and frameworks within which to formulate our moral responsibilities.

Distinguishing ETNL and ETIL

In the event that we find life elsewhere, will we recognize it? Yes. NASA' s Astrobiology Roadmap of 2003 assumes that life has continuity, even if it has one genesis on Earth and a different genesis elsewhere. This assumption permits research aimed at producing a *universal theory of life*, one that applies terrestrially and extraterrestrially. "The origin of life on Earth is likely to represent only one pathway among many along which life can emerge. Thus the universal principles must be understood that underlie not only the origins of life on Earth, but also the possible origins of life elsewhere" (NASA 2003).

This is the first in a long list of assumptions we make that permits us to launch the astrobiological research agenda. We assume that a universal theory of life is possible, even though we do not have such a theory yet. To date, no one can explain the origin of life on Earth. We can explain the evolution of species once common descent has begun; but the origin of life is not included in the Darwinian or neo-Darwinian model of evolution. Yet we have accustomed ourselves to projecting both a second genesis of life as well as evolutionary development onto worlds elsewhere in space.

Now, assuming a continuity of life on Earth and elsewhere, we may draw a line between nonintelligent and intelligent life. More than likely, because of relative proximity, we will find extraterrestrial nonintelligent life (ETNL) forms nearby, within our Solar System. More than likely, if we are to passively receive contact from an extraterrestrial intelligent life (ETIL) form, it will signal us from elsewhere in the Milky Way. This suggests two distinctive domains within which to pursue astroethics.

The search for ETNL looks quite different from the search for ETIL, at least in current practice. The search for ETNL is taking place here within our own Solar System, with a focus on Mars as well as the moons (such as Titan, Europa, Ganymede, and Callisto) of Saturn and Jupiter.[1] SETI (the search for extraterrestrial life), in contrast, aims its radio telescopes at other stars in the Milky Way and occassionally at even more distant galaxies. The exobiological search for ETNL is intrusive; it employs spacecraft, landings, and visits to other possible ecospheres by probes or even astronauts. The SETI search is nonintrusive; it simply listens for signals that might indicate the past or present existence of ETIL. These two different search methods have elicited two corresponding types of ethical reflection.

The search for ETNL is concerned primarily with *planetary protection* (NASA 2010), protection against contamination. The risk of contamination goes in two directions, forward and backward. The possibility of *forward contamination* alerts us to the risk of disturbing an already existing ecosphere; the introduction of Earth's microbes carried by our spacecraft or equipment could be deleterious to an existing habitable environment. *Back contamination* would occur if a returning spacecraft brings rocks or soil samples that contain life-forms not easily integrated into our terrestrial habitat. A quarantine program will be required to determine the safety of newly introduced ETNL.

Margaret Race and Richard Randolph have proposed four underlying principles for developing an ethics appropriate to the discovery of nonintelligent life in our Universe: (1) Cause no harm to Earth, its life, or its diverse ecosystems; (2) respect the extraterrestrial ecosystem and do not substantively or irreparably alter it (or its evolutionary trajectory); (3) follow proper scientific procedures with honesty and integrity during all phases of exploration; and (4) ensure international participation by all interested parties (Race and Randolph 2002; Race 2007, this volume; Sullivan, this volume; see also McKay, this volume, for an alternative perspective).

When it comes to ethical guidelines for ETIL, SETI has already offered a statement: The Declaration of Principles concerning Activities Following the Detection of Extraterrestrial Intelligence. The thrust of

SETI's nine principles is to follow scientific best practices, seek independent confirmation to establish credibility, and announce the discovery only after consultation with international leadership (SETI 1990; Randolph and Race 2002; Race, this volume; Tarter, this volume).

Do we have empirical knowledge that ETIL exists? No. "No unambiguous signals from extraterrestrial intelligence have been detected" (Dick 2003). Yet the search goes on, working with a set of assumptions regarding the likelihood that life has had an independent genesis elsewhere and that any such independent life-form has followed a recognizable path of evolutionary development. If the path of evolution looks like evolution on Earth, then we can project that an extrasolar planet much older than Earth will be home to a much more highly evolved life-form (Tarter, this volume). Such ETIL may have progressed much farther than we in complexity, intelligence, wisdom, science, and technological achievement. Conversely, ET life that began later than life on Earth might still have attained intelligence, but it might be more primitive than *Homo sapiens* here at home. Given such assumptions regarding evolution, we could establish a speculative calculus: planets with a longer evolution time would produce greater rational intelligence, while those with a shorter evolution time would develop a lower level of intelligence. The intelligence level of the ETIL we eventually engage might be measured according to such a calculus.

It must be acknowledged at the outset that the distances in question reduce the likelihood that we will soon find ourselves engaged in a real-time interaction with any extrasolar civilization, let alone one that has not yet developed the technological capability of interstellar communication. Given the limit of the speed of light, and given that such ETIL might be hundreds or thousands of light-years distant, what is more likely is communication that involves considerable delays between question and answer. Even with this acknowledgment of the low likelihood of real-time engagement between Earth and ETIL, we will here try to imagine interactive engagement and try to identify fitting ethical frameworks for developing our moral responsibility.

Astroethical Assumptions for Engaging ETIL

What form might ETIL take? As we have just noted, because no empirical evidence is available, we must speculate on the basis of what we know from the evolution of life on Earth and then be ready for surprises. Might we find intelligence that is silicon based rather than carbon based? Might we find intelligence expressed in entities other than biotic individuals? Might intelligence belong to groups rather than individuals? Such would count as surprises, to be sure. Speculations seem most reasonable when we project the image of an extraterrestrial being who looks somewhat like us, a biotic individual.

These individuals may be more or less advanced on a scale of evolutionary development, astrobiologists surmise. If ETIL forms have evolved longer than we, they might be more complex and more intelligent. If they have evolved less than we, they might be simpler and less intelligent than we. Despite the absence of any empirical evidence, speculations that extrapolate on our understanding of evolution on Earth provide a reasonably coherent matrix within which we might begin ethical deliberation. Coherent astroethical frameworks can be built only upon such speculative foundations.

To limit the scope of application for the ethical analysis I would like to propose, let me set the parameters within a set of assumptions. These assumptions are not arbitrary; they derive from the assumptions within which many of today's astrobiologists are already working.

Assumption 1. We will at some point find ourselves in an interactive engagement with ETIL either on Earth or on the ETIL's home planet. The stage of interactive engagement we are picturing here would come sometime after initial passive contact by SETI.

Assumption 2. Extraterrestrial intelligent creatures will in fact be creatures, intelligent individuals living together in a society. This parameter is warranted because of the working assumption made in the field of astrobiology that life might originate elsewhere in a fashion similar to what happened on Earth, and that a history of evolutionary development parallel to what has happened on Earth might follow. A concomitant assumption is that evolution is progressive—that as time passes, complexity and intelligence advance, and that the degree of intelligence achieved by ETIL would indicate a corresponding span of evolutionary time.

Assumption 3. The level of advance achieved by ETIL would be measured primarily in terms of intelligence. Even though advances in culture, aesthetics, or morality could be relevant to terrestrial ethics, we will limit this discussion solely to intelligence. This restriction is warranted because intelligence is the single category most frequently identified by astrobiologists as a measure of evolutionary progress.

Assumption 4. The level of ETIL intelligence would be measured partly, though not exclusively, by scientific or technological achievement. Many astrobiologists believe that science and technology are what intelligent beings naturally produce: the more highly evolved, the more science and the more advanced in technology. If ETIL exhibit a high level of scientific and technological achievement, we will conclude that they are at least our equals, if not our superiors, in intelligence.

Let me add parenthetically that I do not necessarily subscribe to all of these assumptions; and I acknowledge that many scientists in the field of evolutionary biology deny the doctrine of progress while registering doubt that a second genesis elsewhere could lead to a second development akin to Earth's *Homo sapiens*. Doubt is cast on the assumption that a second evolution would follow the path of progress from simple life-forms to intelligence, because evolutionary biologists find no purpose or direction built into the evolutionary process. What obtains is chance, contingency, or accident, not design. Even intelligence on Earth is an accident, not the product of nature's fine-tuning. Ian Tattersall of the American Museum of Natural History in New York puts it this way: "The human mind is *not* fine tuned for anything. It is the outcome of a whole host of historical accidents" (Tattersall 2006).

For this reason, Ernst Mayr argues that the development of intelligent civilizations with the ability to communicate is so improbable as to render SETI a useless enterprise. Because "evolution never moves in a straight line toward an objective (intelligence)," we cannot expect a repeat of what has happened on Earth somewhere else (Mayr and Sagan 1996). Life as we know it on Earth is unique. Carl Sagan, a contact optimist, counters Mayr by supplying big numbers. Because the number of possible habitable planets is so large, even an improbable event could occur multiple times: "The bulk of the current evidence suggests a vast number of planets distributed through the Milky Way with abundant liquid water stable over lifetimes of billions of years. Some will be suitable for life—our kind of carbon and water life—for billions of years less than Earth, some for billions of years more" (Mayr and Sagan 1996).

It is not my task here to debate the merits of these conflicting positions. My purpose here is to explore ethical principles that might be coherent within the set of very hypothetical assumptions drawn on by contact optimists. As we draw out the ethical implications of the working set of assumptions, these implications may redound with additional doubts about the assumptions on which they are based.

Astroethical Categories for Engaging ETIL

With these assumptions now in mind, I recommend that the next step engage ethical categorization of the moral communities to whom we might become responsible. Once interaction between civilization on Earth and civilization on another planet has commenced, we earthlings will need to name and understand just whom it is we are dealing with. Measured on a scale of intelligence, I suggest we place our newly discovered neighbors into one of three categories: our inferiors, our peers, and our superiors. Given the astrobiological assumptions above, we might very well encounter beings less fully evolved and, hence, less developed in intelligence than we are. We are also likely to meet some intelligent extraterrestrials who approximate our level of evolutionary development; and certainly we might meet some whose evolutionary history is much longer than ours and whose level of achievement is far more complex than ours. Might the framework for deliberating on our moral responsibility depend in part on the category within which our new neighbors belong? Yes.

How might we go about measuring intelligence to determine which of the three ethical frameworks might best apply? The initial stage of engagement will immediately elicit a hypothesis: They are our equals!, They are our inferiors!, or They are our superiors![2] We might then need to test to confirm or disconfirm this initial hypothesis.

Distinguishing Our Peers from Inferior ETIL

How might we distinguish between our peers and our inferiors, especially if we appeal to the criterion of intelligence? Let me suggest three complementary tests—the technology test, the Turing test, and the naming test. The technology test would be quite straightforward: Do our new space neighbors make machines that work as well as ours? As SETI's Seth Shostak puts it: "You're 'intelligent' if you can build a powerful laser or a thumping radio transmitter" (Shostak 2002).

Just how does the alien technology compare to ours? Worse? Better? Do their machines seem to be the product of an intelligence we recognize? Can we learn from the ETIL, and can they learn from us? If we find it to be a two-way street, then perhaps our new space neighbors belong in the peer category. If it turns out to be a one-way street with ETIL as the only learners, then perhaps they belong in the subordinate category.

If the technology test is ambiguous, then we might turn to the Turing test and the naming test. These two tests will depend on the establishment of communication, with back-and-forth conversation. If language is based upon a deep structure, an inherent logic required by its fundamental expression of the relationship of the mind to its physical substrate and interaction with the physical world, then we can have confidence that with work we will be able eventually to understand alien communication. In addition, the philosophical field of hermeneutics describes shared understanding in terms of a fusion of horizons, the slow but sure dialectical construction of a shared horizon of understanding. To proceed with our ETIL speculations, we must operate with the optimistic assumption that conversational bridges are possible.

Our second test, the Turing test, has been applied thus far in measuring the purported intelligence of machines. Proposed by British mathematician Alan M. Turing (1912–1954), the test simply sets up a pattern of blind interaction between a human person and a machine (Danielson 2005). If during the interaction the human perceives the machine partner to exhibit signs of intelligence, then we can conclude that it is intelligent. This is an interactive test. To date no computer has passed the Turing test, despite Internet contests with chatterbots (an automatic program that carries on a conversation). Ray Kurzweil forecasts that computers will "emulate the flexibility, subtlety, and suppleness of human intelligence" sometime after 2020 (Kurzweil 2005). Might we devise a variant on the Turing test to detect the presence of intelligence among ETIL? Or a bit more specifically, to detect a certain feature of intelligence important to our future, namely, the ability to interact? We will need to know what to expect regarding the future of interaction before we can formulate appropriate ethical principles.

Using the naming test, we would ask our new ETIL acquaintances: "What is your name?" If they answer with a name, they will be candidates for peerage. If they have not named themselves, then perhaps they are something less than our peers. The criterion of naming suggests itself from two sources, one theological and the other scientific. Theologically, the naming of the animals provides a significant descriptor of the human being in Genesis 2:19: "So out of the ground the Lord God formed every animal of the field and every bird of the air, and brought them to the man to see what he would call them; and whatever the man called every living creature, that was its name" (NRS).[3] Human beings name animals, not the other way around.

Those among us who live on farms or who have pets know about naming animals. Farmers name breeds. Pet owners name individual dogs or cats. The reverse is not the case. Even though animals possess intelligence and can even interact with us at certain levels, they do not name either themselves or us. We human beings are the sole namers on our planet. Should we engage in conversation with our new space acquaintances and find that they name neither themselves nor us, then the task will be left to us.

Names are important when it comes to engagement. Europeans arriving in the Western Hemisphere named those already living here "Indians." The people we know as the Navajo already had a name for themselves, *Dine*, roughly translated as "the people." As it turns out, Europeans and the *Dine* are intellectual peers. Neither has the right to establish the name for the other. Had the naming test been invoked upon the arrival of Europeans in the New World, perhaps some of history's injustices might not have followed. Learning from this lesson might aid us at the first moment of engagement with extraterrestrial contacts.

Scientific support for the naming test is based on the judgment of many that the emergence of language accompanied by symbolic thinking in biological evolution marks a significant, if not definitive, threshold crossed by our human species. The genes and brains of *Homo sapiens* are much like those of our primate siblings, the chimpanzees, gorillas, orangutans, and bonobos. Yet our cognitive abilities are qualitatively different. Once the doorway to language was unlocked, a coevolution of brain and language ensued. "Language is not merely a mode of communication," writes Terrance Deacon, "it is also the outward expression of an unusual mode of thought—symbolic representation . . . thought does not come innately built in, but develops by internalizing the symbolic process that underlies language. So species that have not acquired the ability to communicate symbolically cannot have acquired the ability to think this way either" (Deacon 1997). If our measure of evolutionary advance is intelligence, then language is the key indicator. "Brain-language co-evolution has significantly restructured cognition from the top-down, so to speak, when compared to other species" (Deacon 1997). The naming test would distinguish a species that has crossed the language threshold from one that has not. After all, said Aristotle, "man is the only animal whom [nature] has endowed with the gift of speech" (McKeon 1941).

If these three tests indicate that the ETIL in question are our inferiors in intelligence, we would then ask: Might the ethical framework for discerning our responsibility toward them be analogous to our responsibility toward Earth's animals? If we answer affirmatively, then we would find ourselves in a classic dialectic. On the one hand, humans exploit all other life-forms, both plants and animals, for human welfare. Animals provide food, work, clothing, and even company. Animals can be sacrificed in medical research to develop therapies that will benefit only human persons. On the other hand, we human beings have a sense of responsibility toward the welfare of animals. We respect them as intelligent beings, and we are concerned about preventing suffering to animals. In some instances, we exert considerable energy and effort to preserve their species from extinction and to ensure the health of individual animals. We love our pets to a degree that rivals loving our own family. In sum we have inherited this double relationship to our inferiors already here on Earth.

This inherited dialectical relationship can be described as conflictual. "Every area of human-animal interaction, be it agriculture, research, hunting, trapping, circuses, rodeos, zoos, horse and dog racing, product extraction, and even companion animals, is fraught with ethical and welfare issues" (Rollin 2005). In early human history, which included the domestication of animals, "husbandry" became the dominating ethic. This "fair and ancient contract" embraced the double principle that animals become better off because humans care for them and protect their welfare, and that humans become better off because they benefit from animal products. Animal ethics included proscriptions against intentional mistreatment or deliberate cruelty. As a consequence humans and animals have coevolved. In the post–World War II era, however, the industrialized production of animal products has removed the "welfare" component. Animal suffering has become massive: confinement, loneliness, boredom, and disease, ending in the terrors of the slaughterhouse. This suffering is not due to intended cruelty but due to the efficiency called for by industrial production. The previous husbandry ethic is being replaced with an "animal rights" ethic as our society grapples with the reinstitution of the "fair and ancient contract." Columbia University's Richard Bulliet forecasts that "the future of human-animal relations in real-world terms will be determined by the worldwide expansion of exploitation in a late domestic mode and the reaction to that expansion by increasingly angry post-domestic activists" (Bulliet 2005).

Important for our application to the upcoming ETIL question is the ethical criterion for rendering moral policy. Is the animal rational in the same sense that we *Homo sapiens* are rational? No, says the tradition beginning with Aristotle. The human is *the* distinctively rational animal.

The separation of humanity from the animal world due to a separation on the scale of rational intelligence justifies an ethic whereby the superior human exploits the inferior animal. "Irrational animals are natural slaves, and no positive human moral or political categories can govern humankind's relations with them" (Fellenz 2005). Yet we might ask: Is rational intelligence the only ethical criterion? No, say some moralists. By extension our moral commitment to protect humans from suffering should be applied to animals. How an animal feels is morally significant. "Animal welfare is most crucially a matter of the animal's subjective experience—how the animal feels, whether it is in pain or suffering in any way" (Rollin 2005; see also Cleland and Wilson, this volume).

With this experience of relating to animals on Earth in mind, our first ethical question in the case of engagement with inferior ETIL would be: By which criterion do we orient our ethics, ETIL's relative rational intelligence or our responsibility for the welfare of ETIL? Our second ethical question would be: Which of these two habits should we invoke: exploitation of ETIL for our own use, or protection of ETIL from suffering? More than likely, all of these alternatives would inform the policies we develop. In ethical categories, we would not impute dignity to ETIL whose level of rational advance falls significantly short of our own. We very well might show them respect, even care. Motivated by faith, we might view ETIL as creatures precious in the eyes of God. In economic categories, more than likely we would exploit less-intelligent ETIL for increased terrestrial prosperity. We might work out terms of exchange or, more likely, simply set up an infrastructure for ongoing exploitation. Would we exploit with moral abandon? Or would we exploit only to the limit set at the point of detriment to the welfare of the ETIL themselves? This raises the matter of our moral responsibility. In terms of our responsibility, I believe we should take the initiative to extend concern for the welfare of such ETIL on the model of our current concern for the subjective quality of animal experience. We should do what we are able to protect ETIL from suffering and enhance their experience of well-being.

Ethics for Engaging Peer ETIL

If we conclude that ETIL are our peers in rational intelligence, then we might find appropriate the golden rule and Immanuel Kant's categorical imperative. Jesus's version of the golden rule is familiar to us all: "In everything do to others as you would have them do to you" (Matthew 7:12, NRS).

For Kant, the formal principle from which all moral duties are derived is this: "I ought never to act except in such a way *that I also will that my maxim should become a universal one*" (Kant 1948; Narveson 1985).[4] In sum we should treat peers as equal to ourselves; and we should care for their welfare just as we would care for our own (see Stoeger, this volume; Sullivan, this volume; Cleland and Wilson, this volume).

Critics of the golden rule say it should not apply to space aliens. Why? Because ETIL might differ from us. Therefore, we should treat space aliens not as we would treat ourselves but, rather, in the way they wish to be treated ("Religion: Space Ethics" 1956). Such an argument can be dismissed as a tautology. The way we want to be treated along with the way they want to be treated is just what the golden rule targets.

Jesus's golden rule and Kant's categorical imperative have greatly influenced the value system of the Enlightenment and, hence, the modern culture of which we are a part. If we find that ETIL resemble us enough to be considered our peers, then we might invoke the value system of the Enlightenment—that is, we might invoke the golden rule and impute dignity to our space neighbors, honoring the way they declare they wish to be treated. Our moral disposition would be to approach our new neighbors with operative values such as equality, liberty, dignity, justice, and mutuality. When it comes to dealing with ETIL as individuals, we would impute dignity to them—that is, we would treat each as a moral end and not merely as a means. "Act in such a way that you always treat humanity, whether in your own person or in the person of any other, never simply as a means, but always at the same time as an end" (Kant 1948). In a more contemporary and theological medium, Alan Falconer expresses it this way: "Human dignity is the inherent worth or value of a human person from which no one or nothing may detract" (quoted in Childress and Macquarrie 1986). Might we impute dignity to ETIL? Might we treat them with dignity? Might "dignity" become the label for identifying our responsibility?

Dignity comes in both theological and secular forms. Theologically, dignity is the gift of God. The "gospel in miniature," John 3:16, reports that "God so loved the world (*kosmos*), that he gave his only begotten Son." God's act of redemption in Jesus Christ confers dignity on the world, and on us. God's love for the world elicits in us a sense of value or worth. God confers upon us dignity, and we in turn affirm our dignity and claim that dignity. God's conferring of dignity upon us is an act of divine grace; it is not based on any quality we ourselves possess. The more secular variant of dignity, which we find in Kant, who inherited it from Greek philosophy, associates dignity with our rational capacity. Rational intelligence warrants being treated with dignity. If we lack rational capacity, we cannot claim dignity. These two variants—dignity due to rational capacity or dignity due to gracious conferral regardless of rational capacity or any other capacity—can cohere as long as we posit that every individual human being belongs in the set or class of rational human beings. But what happens when a given individual does not meet the rationality criterion? Does he or she lose dignity? Not exactly. Infants and elderly people suffering from dementia may lack rational capacity; yet we treat them with dignity because they are human. We confer dignity upon less than fully rational members of the human species. Evidently, belonging to a class of rational beings trumps exhibiting the quality of rationality by each individual within the class. Based upon these post-Christian Enlightenment values, every human person on Earth can claim the right to be treated with dignity.

What might this imply as we construct ethical policy for engaging peer ETIL? Perhaps we could adapt portions of the *Universal Declaration of Human Rights,* passed by the General Assembly of United Nations in 1948. We would rewrite them to refer to the ETIL we have deemed to be our peers. Where we read "human beings," we could substitute the name that the ETIL apply to themselves.

Article 1: All human beings are born free and equal in dignity and rights. They are endowed with reason and conscience and should act toward one another but in a spirit of brotherhood.
Article 3: Everyone has the right to life, liberty, and security of person.
Article 5: No one shall be subjected to torture or to cruel, inhuman, or degrading treatment or punishment.

The imputation of dignity toward ETIL should be accompanied by a denial of our own right to unilaterally exploit them. We might encourage the development of bilateral commerce, of course; but we should do so presuming the equality and liberty of our trading partners. We might also restrict our intrusion into their ecosphere. We might adapt for ETIL the Race and Randolph principle aimed at ETNL: "Respect the extraterrestrial ecosystem and do not substantively or irreparably alter it (or its evolutionary trajectory)." In sum, the ethical principles we invoke to deal with peer ETIL might draw upon our Enlightenment values. We could formulate to make applicable to ETIL the principles that we now invoke to maintain terrestrial justice and peace.

Refining the Categories

Before proceeding to ethical analysis of our third category for identifying relevant moral communities, superior ETIL, we might wish to refine the three basic categories. I suggest we divide peer ETILs into two subcategories: hostile and peaceful; and I suggest we divide the superior ETILs into three subcategories: hostile, peaceful, and salvific. Once we have discerned that ETIL are our equals or our superiors in technology and perhaps in intelligence, we will need to ask whether or not they pose a threat to Earth's security and well-being. How we answer this question may partially guide the moral direction we take.

The anxiety associated with insecurity leads us *Homo sapiens* to strike out with violence (Peters 1994). We on Earth will find ourselves uneasy, on the verge of violence, until we can be assured that the ETIL we confront mean us no harm. Whether the high-minded among us find it moral or not, the reality is that no rational discourse about ethics can take place when anxiety is high and security is low. To determine whether ETIL are a threat or not will inescapably become our first priority.

In the event that the ETIL in question are hostile, we will find ourselves working within a framework that includes both the imputation of dignity and our pressing need to protect our planet from alien exploitation or damage. We know from experience that whenever we are confronted with a hostile enemy from without, we find ourselves within our society compromising human dignity. Our political leaders try to persuade our society that our targeted enemies should be reduced to "inhuman," if not demonic, status, and this justifies going to war. What this indicates is that the social psychology of self-defense pits human dignity against the mustering of military support. Security trumps dignity. If threatened by alien hostility, we can forecast that military rhetoric will attempt an equivalent of dehumanizing the ETIL and hence denying that they have dignity. A nation's leaders simply cannot embrace Jesus's peace ethic of loving our enemies combined with turning the other cheek (Matthew 5–7). So, as difficult as it may sound, we will need an ethic that affirms the dignity of ETIL while rallying our Earth allies in planetary defense. We might need to adapt for peer ETIL the Race and Randolph principle, "Cause no harm to Earth, its life, or its diverse ecosystems," within a tense relationship to the wider ethical principle of imputing dignity to our extraterrestrial peers.

In the event that peer ETIL prove to be neutrally peaceful or even benevolent, then the principles giving expression to Enlightenment values should prevail without challenge: equality, liberty, dignity, and mutuality.

Ethics for Engaging Superior ETIL

It is difficult to *imagine up*. It is easier to *imagine down*. When comparing humans with animals, for example, we can imagine down by distinguishing things we humans can do that are beyond the capability of our animal neighbors. When is comes to imagining ETIL who might be superior to us in intelligence, it is difficult to imagine what superior intelligence would manifest that is beyond the very human intelligence that is doing the imagining. This puts initial constraints or limits on how we can begin to approach the topic of ethics when engaging ETIL more advanced than Earth's *Homo sapiens*. Nevertheless, it is incumbent in astrobiological ethics to speculate about the possibility of engaging with intelligent beings who are superior to us.

If we meet ETIL superior to ourselves, will they be hostile, neutrally peaceful, or salvific? Given the assumptions made by astrobiologists that extraterrestrial evolution will follow a path toward increased intelligence, as it has on Earth, the prospect of ETIL fitting the hostile category is to be expected. Charles Darwin's key evolutionary principle is "natural selection," which he identifies with "the struggle for existence" and with Herbert Spencer's phrase "survival of the fittest" (Darwin 2003). In the struggle for existence, living creatures undergo cruelty, suffering, and waste (Darwin 2003). And the species to which virtually every individual creature belongs will eventually go extinct to make way for a more-fit species. The strong devour the weak. The big eat the small. The fit survive in a world that is, as Tennyson put it, blood "red in tooth and claw."

Even slavery is not unknown in the prehuman struggle for existence. One species of red ant (*Formica sanguinea*) enslaves a species of black ants (*Formica fusca*) to gather food and make nests, observes Darwin. "The slaves are black and not above half the size of their red masters" (Darwin 2003). If we would substitute intelligence for size and ascribe it to the ETIL we meet, perhaps they might decide to enslave *Homo sapiens* for their gain in the struggle for existence. This prospect would be most consistent with the theory of evolution as we project it onto planets among the stars.

Given astrobiological assumptions regarding a repeat of evolution on extraterrestrial planets, hostility is what we should expect from ETIL. Yet surprisingly, some SETI speculators anticipate meeting intellectually superior ETIL who will benevolently help us on Earth. For this reason, I add the subcategory of *salvific*.

How do we get from the struggle for existence to extraterrestrial saviors? How does evolution transcend itself?

We are trying to discern the logic inherent within astrobiological thinking. Some in the astrobiology community project an image of a more highly evolved extraterrestrial creature, who would like to rescue us earthlings from the ignorant habits we have developed due to our inferior level of intelligence. Because we on Earth have not yet achieved the level of rationality necessary to see that international war and planetary degradation are inescapably self-destructive, we could learn from ETIL who are more advanced than we. Such thinking is obviously myth, not science. No empirical evidence justifies such speculation; yet such dreaming of redemption descending from the skies is tantalizing to the terrestrial imagination. I have noted elsewhere that included in much of astrobiological theorizing is a version of the *ETI myth* (Peters 2008, 2009). The essence of the ETI myth is that science saves: science can save Earth from its inadequacies, its evolutionary backwardness, and its propensity for self-destruction. And if terrestrial science is insufficient to save Earth, then extraterrestrial science just might be sufficient.

By *myth* here I refer to a cultural construct, a perceptual set, a window frame, so to speak, through which we look in order to view the world out there. At work in modern culture in general, as well as in astrobiology in particular, is an identifiable framework—a myth, if you will—within which we cast the questions we pose to the mysteries evoked by our experience with outer space. The ETI myth begins to reveal its shape as SETI's Frank Drake gives voice to speculations:

> Everything we know says there are other civilizations out there to be found. The discovery of such civilizations would enrich our civilization with valuable information about science, technology, and sociology. This information could directly improve our abilities to conserve and to deal with sociological problems—poverty for example. Cheap energy is another potential benefit of discovery, as are advancements in medicine.[5]

Note the optimism. Drake does not expect what Darwin would expect, namely, an extraterrestrial race, engaged in the struggle for existence, that might like to exploit us on Earth. Rather, Drake's extrapolation of evolution to ETIL imagines an intelligent and benevolent race ready to offer us aid and assistance. His vision includes optimism regarding the solution to Earth's "sociological" problems such as poverty and energy. Space visitors might even give us a leap forward in medicine.

What Drake believes is that science is salvific, and extraterrestrial science would be even more salvific than terrestrial science. Should an extraterrestrial civilization more evolutionarily advanced than us engage planet Earth, we could benefit from the ability of ETIL to save us from our own primitive inadequacies and even our own propensity for self-destruction. It is this thought structure within astrobiology that warrants the designation for more highly evolved and more intelligent ETIL as "salvific."

For ETIL to become salvific for us on Earth, ETIL would necessarily be altruistic. By *altruism* we refer to a motivational disposition to aim conduct toward "the good of the other person" (Latin *alter*, "the other") (Childress and Macquarrie 1986). We would be "the other" in the eyes of superior ETIL. What warrants our expecting altruism from evolving ETIL? An assumption with two overlapping applications is at work here. The first is this: To be more highly evolved is to be better. The second is this: Beings who are more highly evolved than we are will have developed not only science but also altruistic virtue. A century and a half ago, social Darwinist Herbert Spencer enunciated the core assumption this way: "The conduct to which we apply the name good is the relatively more evolved conduct; and . . . bad is the name we apply to conduct which is relatively less evolved" (Spencer 1879). With such an assumption, Drake and others can speculate that a more highly evolved race of intelligent beings will have surpassed us not only in science but also in morality. These advances will place ETIL in a position of saving us on Earth from our backwardness and from the dangers of ecological or thermonuclear self-destruction. What is happening here? The idea of salvific power on the part of a more advanced extraterrestrial life-form is the projection of Earth's astrobiologists who already see terrestrial science as potentially salvific. A more advanced extraterrestrial version of science combined with altruistic virtue prompts in us a vision of celestial science descending from the sky to rescue us from our primitive and still self-destructive human propensities.

In short we find within space speculation an image of evolutionary perfection that is inconsistent with what the more conservative Darwinian model would predict. Given Darwin's description of the "struggle for existence," one would expect more highly evolved ETIL to exploit us, perhaps even enslave us. Drake's SETI mindset, however, evoking the optimism of social Darwinism, predicts an advanced stage in evolution where the principles of past evolution would no longer apply, where ETIL would be so altruistic as to desire to aid us on Earth without expecting reciprocity. Evolution will have produced celestial saviors coming to our planet to rescue us from our own underdeveloped intelligence.

This predictive conflict is reminiscent of Richard Dawkins, who, on the one hand, says that the entire history of evolution has been driven by the "selfish gene," but also, on the other hand, says that we in the modern world can overcome our genes and adopt liberal ethics. "We have the power to defy the selfish genes of our birth and . . . cultivating and nurturing pure, disinterested altruism—something that has no place in nature" (Dawkins 1989). Dawkins articulates what seems to be assumed by some SETI voices: even though evolution to date has been cruel and selfish and destructive, eventually with more highly evolved intelligence creatures will become so altruistic as to leave their evolutionary background behind. One might ask how such a prediction could be based upon evolutionary precedents—how this evolutionary leopard could change its spots. Yet we can find within SETI a prediction of scientific salvation from ETIL. So, I will ask here about the ethical implications of such an optimistic prediction right along with predictions based on the more standard Darwinian model.

If superior ETIL follow the Darwinian model and confront us with hostile and exploitative enslavement, then perhaps we will frame our ethics accordingly. The New Testament provides instructions for slaves: "Slaves, accept the authority of your masters with all deference, not only those who are kind and gentle but also those who are harsh" (1 Peter 2:18, NRS). This may seem unrecognizable. The treatment of the superior master by an inferior slave has fallen into disuse in our post-Enlightenment period. This is because of the erasure of the line between superior and inferior human beings within modern Enlightenment culture. We are all equal—that is, we are all ethically equal. Each of us has dignity by virtue of our belonging to the human species, and slavery violates the principle of dignity. Should a master–slave relationship rear its ugly head somewhere on our planet, we children of the Enlightenment would encourage the slaves to rebel and strive for their own liberation. Such a moral commitment to liberation would be justified by the assumption that masters and slaves are equal.

When we use the assumptions made by many in the astrobiology field, however, we cannot coherently make the argument that all intelligent beings are equal. Those who have evolved longer and who have attained a higher level of rational intelligence would be, by definition, superior to us. We could not justify liberating ourselves from their rule with an argument based upon equality. If we allow a New Testament influence, we might consider developing an ethic of slave responsibility, even loyalty, if not love, for our masters.

Might loyal slave morality have its limits? What if our superior masters request or even demand of us slaves something we deem immoral? Jan

Narveson poses this question. "If another set of beings, invulnerable to our best efforts to destroy, damage, or injure them and overwhelmingly more powerful than us in every way, threatened us with various kinds of coercion which they were perfectly able to apply unless we did their bidding, should a moral human being do the bidding in question? I think it would be hard to give a general answer. It would surely depend on what they wanted us to do" (Narveson 1985).

What might we terrestrials think if we are asked to make destructive weapons? Let us consider an analogy. Currently in the United States, selected companies design and manufacture increasingly sophisticated and deadly weapons, which are sold for use first to the U.S. military, then to governments friendly to the United States, then to terrorists and drug cartels. Those who work in this industry contribute to both widespread death and big profits. The moral conscience of some workers causes them to avoid such employment. Because an enslaving ETI group would need weapons to maintain its hegemony, we Earthlings might be asked to provide support for such a rule. Might we confront a moral dilemma between our responsibilities as slaves and our responsibilities to our own moral judgment?

In the event that ETIL approach the civilizations on Earth in a peaceful manner, we would want to respond with an appropriate ethic. Maintaining peace would become an immediate moral commitment. We might even find ourselves organizing to quiet down and restrict earthly voices that would disturb the peace. We would want to police ourselves in the name of peace. Peace would benefit life on Earth. In addition, moral policies we set would likely treat our ETI beings with dignity, respect, and courtesy due to their position of superiority and potential power.

In the event that ETIL turn out to be not only more intelligent but also altruistic toward us, then an ethic of gratitude might be included in our responsibility. We would receive and make use of the gifts that increased intelligence would allegedly provide us: such as the means for maintaining a healthy planetary ecology, improvements in our medical care, and more justice in our social practices. Then, we would build upon what we have already said about maintaining terrestrial peace and treating our superiors with dignity; we would add a measure of grateful respect.

In sum we should treat superior ETILs with dignity, respecting and even caring for their welfare. If they are hostile and enslave us, we should invoke an appropriate slave morality that maintains their dignity. If ETIL are peaceful toward us and open up avenues of conversation and commerce, then the principles of justice and the striving to maintain peace

Table 11.1 Ethics for Inferior, Peer, and Superior ETIL

	Inferior ETIL	*Peer ETIL Hostile*	*Peer ETIL Peaceful*	*Superior ETIL Hostile*	*Superior ETIL Peaceful*	*Superior ETIL Salvific*
Ethics?	Respect	Dignity	Dignity	Dignity	Dignity	Dignity
Economics?	Exploitation	Exploitation	Commerce	Commerce	Commerce	Commerce
Responsibility?	Care	Defense	Justice	Slave Ethics	Peace	Gratitude

should obtain. If out of their superior wisdom and altruistic motives ETIL seek to better our life here on Earth, we should accept the gifts they bring and respond with an attitude of gratitude.

Table 11.1 summarizes some of the options for earthling interactions with ETIL, based on the relative status of ETIL.

Conclusion

Astroethics today is necessarily a speculative endeavor. The astrobiology upon which astroethics deliberates is itself speculative. When it comes to extraterrestrial intelligent life-forms, terrestrial scientists are comfortable thinking of extraterrestrial habitats as harboring a separate genesis of life and a story of evolution parallel to Earth's story. Evolution in this case is assumed to be progressive, following an entelechy toward increased rational intelligence. In the case where the length of evolutionary development is less than or comparable to our own, we can expect inferior or equal levels of rational capacity. In the possible case where an extraterrestrial race has had more time to evolve, we can expect a level of rational intelligence superior to our own. Astroethicists should be ready to construct a framework for moral responsibility that corresponds to these three relevant kinds of moral communities.

Notes

1. "Observations of Saturn's moon Titan over several years show that its rotational period is changing and is different from its orbital period." What this most likely means is that Titan's shifting crust sits atop an interior ocean, perhaps a water-ammonia ocean that is "potentially habitable. Internal oceans have been detected magnetically on Europa and Callisto and are also presumed to exist on Ganymede" (Lorenz et al. 2008, pp. 1649, 1650).

2. The three parallel yet not identical categories used by Jan Narveson are "superhuman," "subhuman," and "very different nonhumans" (Narveson 1985).

3. "NRS" refers to the New Revised Standard Version, the English translation of the Bible released in the United States in 1989.

4. Might we expect the extraterrestrials to be moral beings? Michael Ruse gives a positive answer to this on the grounds that morality is adaptive and highly evolved extraterrestrial beings will have discovered this just as we earthlings have. "What I argue, therefore, is that if, indeed, natural selection is at work on our extraterrestrials, we might expect to find something akin to the Categorical Imperative having evolved amongst them" (Ruse 1985, 65).

5. Quotation from Richards 2003.

References

Arnould, J. 2005. "The Emergence of the Ethics of Space: The Case of the French Space Agency." *Futures* 37: 245–254.

Bulliet, R. W. 2005. *Hunters, Herders, and Hamburgers: The Past and Future of Human–Animal Relationships.* New York: Columbia University Press.

Childress, J. F., and J. Macquarrie, eds. 1986. *The Westminster Dictionary of Christian Ethics.* Louisville, KY: Westminster John Knox Press.

Danielson, P. 2005. "Turing Tests." In *Encyclopedia of Science, Technology, and Ethics,* edited by Carl Mitcham. 4 vols. New York: Macmillan, Thomson Gale.

Darling, D. 2001. *Life Everywhere: The Maverick Science of Astrobiology.* New York: Basic Books.

Darwin, C. 2003. *The Origin of Species by Means of Natural Selection of the Preservation of Favoured Races in the Struggle for Life.* 6th ed. New York: Signet.

Dawkins, R. 1989. *The Selfish Gene.* Oxford: Oxford University Press.

Deacon, T. W. 1997. *The Symbolic Species: The Co-evolution of Language and the Brain.* New York: W. W. Norton.

Dick, S. J. 2003. "Extraterrestrial Life." In *Encyclopedia of Science and Religion,* edited by J. Wentzel and V. van Huyssteen. New York: Macmillan: Thomson/Gale.

Fellenz, M. R. 2005. "Animal Rights." In *Encyclopedia of Science, Technology, and Ethics,* edited by C. Mitcham. New York: Macmillan, Thomson/Gale.

Jonsen, A. R. 1986. "Responsibility" In *The Westminster Dictionary of Christian Ethics,* edited by J. F. Childress and J. Macquarrie. Louisville, KY: Westminster John Knox Press.

Kant, I. 1948. *Groundwork of the Metaphysic of Morals.* Translated by H. J. Paton. New York: Harper.

Kurzweil, R. 2005. *The Singularity Is Near: When Humans Transcend Biology.* New York: Penguin.

Lorenz, R. D., B. W. Stiles, R. L. Kirk, et al. 2008. "Titan's Rotation Reveals an Internal Ocean and Changing Zonal Winds." *Science* 319: 1649–1651.

Mayr, E., and C. Sagan. 1996. "The Search for Extraterrestrial Intelligence." *Planetary Report* 16: 4–13.

McKeon, R., ed. 1941. *The Basic Works of Aristotle.* New York: Random House.

——. 2010. "Planetary Protection." http://planetaryprotection.nasa.gov/.

Narveson, J. 1985. "Martians and Morals: How to Treat an Alien." In *Extraterrestrials: Science and Alien Intelligence*, edited by E. Regis Jr. Cambridge: Cambridge University Press.

NASA. 2003. *Astrobiology Roadmap*. Moffat Field, CA: NASA Ames Research Center. http://astrobiology.arc.nasa.gov/roadmap/roadmap.pdf.

Peters, T. 1994. *Sin: Radical Evil in Soul and Society*. Grand Rapids, MI: Eerdmans.

———. 2003. *Science, Theology, and Ethics*. Aldershot, UK: Ashgate.

———. 2008. *The Evolution of Terrestrial and Extraterrestrial Life*. Goshen, IN: Pandora Press.

———. 2009. "Astrotheology and the ETI Myth." *Theology and Science* 7: 3–30.

Race, M. 2007. "Societal and Ethical Concerns." In *Planets and Life: The Emerging Science of Astrobiology*, edited by W. T. Sullivan III and J. A. Baross. Cambridge: Cambridge University Press.

Race, M. S., and R. O. Randolph. 2002. "The Need for Operating Guidelines and a Decision Making Framework Applicable to the Discovery of Non-Intelligent Extraterrestrial Life." *Advances in Space Research* 30: 1583–1591. www.seti.org/pdfs/m_race_guidelines.pdf.

Randolph, R. O., M. S. Race, and C. P. McKay. 1997. "Reconsidering the Theological and Ethical Implications of Extraterrestrial Life." *CTNS Bulletin*, 17: 1–8. www.seti.org/pdfs/m_race_ethics.pdf.

"Religion: Space Ethics." 1956. *Time*, October 1. www.time.com/time/magazine/article/0,9171,862394,00.html.

Richards, D. 2003. "Interview with Dr. Frank Drake." *SETI Institute News* 12: 5.

Rollin, B. E. 2005. "Animal Welfare." In *Encyclopedia of Science, Technology, and Ethics*, edited by C. Mitcham. New York: Macmillan, Thomson/Gale.

Ruse, M. 1985. "Is Rape Wrong on Andromeda?: An Introduction to Extraterrestrial Evolution, Science, and Morality." In *Extraterrestrials: Science and Alien Intelligence*, edited by E. Regis Jr. Cambridge: Cambridge University Press.

SETI. 1990. "The Declaration of Principles concerning Activities Following Detection of Extraterrestrial Life." *Acta Astronomica* 21: 153–154. www.setv.org/online_mss/SETI-DofP90.pdf .

Shostak, S. 2002. "Are We the Galaxy's Dumbest Civilization?" Accessed May 6, 2008. www.seti.org/epo/news/features/are-we-the-galaxys-dumbest.php.

Spencer, H. 1879. *The Data of Ethics*. New York: A. L. Burt.

Tattersall, I. 2006. "The Origins of Human Cognition and the Evolution of Rationality." In *The Evolution of Rationality: Interdisciplinary Essays in Honor of J. Wentzel van Huyssteen*, edited by F. L. Shults. Grand Rapids, MI: Eerdmans.

United Nations. 1948. *Universal Declaration of Human Rights*. www.un.org/Overview/rights.html.

CHAPTER TWELVE

A Scientifically Minded Citizenry

The Ethical Responsibility of All Scientists

Erika Offerdahl
NORTH DAKOTA STATE UNIVERSITY

What is the purpose of education? For thousands of years, societies, teachers, and students have wrestled with this question. Today the United States is trying to upgrade education, and particularly its science, technology, engineering, and mathematics education, to create not only future scientists and engineers but a more literate general population. In addition to this push for science literacy, some educators propose both to regenerate the ideal that education not only imparts information but also promotes character development, and to improve content knowledge and skill while increasing citizens' understanding of the importance of education and science to the larger community (Roth 2012). Astrobiology provides a unique opportunity to generate interest in the study of science and explore the interdependence of scientific advances with individual and societal philosophical foundations.

For over 50 years, scientific literacy has been a laudable goal of science education. While there is variation in what is meant by scientific literacy, there is general consensus about its benefit to society. In other words, although only a small fraction of the population will become scientists, it is thought to be ethically responsible to foster a society with the skills necessary to understand the role and implications of scientific discovery. The potential to raise the level of scientific literacy within our own country is increasing, as a greater number of people from diverse backgrounds are taking advantage of postsecondary education (U.S. Census Bureau 2009). Yet there is mounting evidence that traditional, lecture-only teaching methods in higher education are hit-and-miss at developing scientific understanding (Feinstein 2011; Handelsman et al. 2004; NRC 2000).

Efforts to increase scientific literacy are further complicated by the nature of modern science, which is generating new knowledge at an exponentially increasing pace. The ways in which we think about and engage in science are rapidly evolving (AAAS 2011). Global access to large, complex data sets produced through interdisciplinary collaborations is becoming the norm. This evolution in our science has profound implications for science education. Preparing future citizens and scientists alike will prove ever more challenging as the requisite set of factual knowledge necessary for even the most rudimentary level of scientific literacy continues to grow. After decades of inattention to the numerous calls for science education reform (see, for example, AAAS 1989, NRC 2003a, 2003b), we stand at a crossroads. We can continue to merely *provide instruction*, despite decades of evidence demonstrating the ineffectiveness of our methods, or we can do what is right. We can choose instead to transform the ways in which we teach to *produce learning* (Barr and Tagg 1995).

What Is Scientific Literacy?

Although the general idea that the public should have some basic knowledge of science goes as far back as the turn of the twentieth century (Shamos 1995), the term "scientific literacy" was likely first coined in the late 1950s as a result of the space race between the Soviet Union and United States (Hurd 1958). This technological rivalry became an important element of the cultural and ideological conflict between these two nations during the Cold War (circa 1947–1991). The Soviets' early success, and America's initial failures, in launching satellites into orbit had implications for public morale and feelings of national security in the United States; a Soviet satellite in orbit projected the ability to target nuclear bombs at the United States. Drawing from the assumption that progress in science depends to a large extent on public understanding and support of scientific research, the American science community sought mechanisms to increase scientific literacy as one strategy for combating the perceived Soviet threat and for harnessing public support for scientific research and development. The result was a push for science education reform on a scale never before seen in the United States (DeBoer 2000). And along with it came an evolving debate about who should be the target audience of such reform efforts and what constitutes scientific literacy (Laugksh 2000).

In the ensuing decades, scientific literacy has been almost universally accepted as a desirable goal for all nations aspiring to economic growth. Yet the arguments in support of this movement are far from coherent. Some advocates posit that the economic health of a nation is tightly tied to the literacy of its citizens. Implicit in this argument is the assumption that competitiveness in the global arena requires the ability to develop and sustain a major presence in the market of high-technology products. This, in turn, relies on a healthy national program for scientific research and development, the cornerstone of which is a scientifically educated populace. In contrast, others advocate the importance of scientific literacy from the standpoint of the individual. The assumption is not only that a more literate population would be more likely to support a national research program, but that such a population would be better equipped to understand and make decisions about the science- and technology-related aspects of their lives, leading to improved quality of life. The emphasis here is not on the betterment of the nation, but on that of the individual.

Policy makers and educators have attempted to deliver on the goal of public science literacy. These efforts often start with definitions, which appear in a wide range of science education reform documents, such as *Beyond 2000: Science Education for the Future and Science for All Americans* (AAAS 1989, 1993; Millar and Osborne 1998). Although there is no single consensus on what it means to be scientifically literate, there are certainly common elements that revolve around themes such as understanding the nature of science, the relationship between science and society, the theoretical underpinnings of the "core" sciences (biology, physics, chemistry), and a basic level of science content understanding in these sciences.

Variability in conceptions of scientific literacy seems to depend on the anticipated end-products of instruction and other efforts toward increased literacy. For example, the definition of scientific literacy might be significantly different for those who envision the end-product to be an individual's ability to read the newspaper and have some basic understanding of the implications of science and technology news for their daily lives, compared to those who see the end-product as the ability to critically evaluate and respond to science and technology news. This variability is visible not only in the curricular choices and teaching practices within the formal education system, but also in how science is communicated to the public through informal science and public outreach efforts.

Establishing consensus about science literacy and what one should know to be considered scientifically literate will be further complicated as the range of questions answerable by science expands. Astrobiologists strive to understand the origins and distribution of life on Earth and beyond. Astrobiology is intrinsically interdisciplinary, so understanding its linked successes necessarily involves concepts in earth, space, physical, and life sciences. The scientific knowledge resulting from this quest will have significant implications, not only for our ways of life, but also for how we understand our place in the Universe. It is not yet clear whether a greater level of scientific literacy will be required than for previous generations. But what is certain is that citizens will be faced with ever more complex questions about the relationship between scientific advancements and their philosophical and spiritual identities—questions not previously considered in terms of science literacy.

What Does the Research Tell Us about How to Teach?

For as long as scientific literacy has been a goal of science education, there have been national documents with long lists of recommendations for how to achieve it (for example, U.S. Department of Education 1983; NRC 1996). Historically, the majority of these documents focused on how science should be taught in the primary and secondary schools (K–12). But increasing access to and diversity in higher education has led to a number of national calls for changes in how we teach science at the undergraduate level (AAAS 2011; NRC 2003a). There is agreement that the undergraduate curriculum should include goals for developing students' understanding of the nature of science, in addition to providing a deep understanding of basic science content.

Developing scientific literacy requires an understanding of the nature of science. Historically science has most often been portrayed as a static body of knowledge to be acquired by students. In contrast, students should learn that science is uncertain, subject to sociocultural influence, and often unable to produce clear results. They should be engaged in activities that mimic the ways in which scientists generate and validate new knowledge (AAAS 2011; NRC 2000). Within the K–12 literature, there is evidence that instructional practices and curricula with explicit learning goals about science as a process are effective at developing students' ideas about the nature of science (see, for instance, Khishfe and Abd-El-Khalick 2002; Lederman 1986; Zeidler and Lederman 1989). But most undergraduate curricula lack such well-articulated learning goals. Moreover, the structure of undergraduate curricula and courses tends to compartmentalize science into discrete disciplines that focus on particular questions rather than an integrated, interdisciplinary way of understanding the world, let alone any discussion of the societal implications of the science. As a result, students become entrenched in the nuances of the basic disciplines, and develop an incoherent understanding of the nature of science in general (Parker et al. 2008). There is evidence that this phenomenon is not limited to general students, but also impacts students who actively pursue an academic pathway in the sciences (Miller et al. 2010). Often knowledge in students' science courses is depicted as universal and objective. Students majoring in science may be exposed to such epistemic views for longer than students majoring in the humanities, thereby reinforcing these incoherent views of the nature of science (Liu and Tsai 2008, Parker et al. 2008).

For many, scientific literacy implies that in addition to understanding science as a process, a person will have achieved some minimum level of content knowledge. Often this gets translated through formal science curricula as a checklist of facts that students should have acquired rather than a conceptual understanding reflecting big ideas and themes in science. At the undergraduate level, students are commonly still taught about science as discrete topics separate from one another, thereby perpetuating students' tendencies to compartmentalize their science understanding. Even in "general education" courses such as astronomy, geology, and environmental science, where the goal is to round out the education of university students and increase literacy, the end result is generally a piecemeal understanding strictly limited to the course syllabus with no connections to real-life applications or broader philosophies. Textbooks contribute to this problem, because they are overstuffed and filled with facts and jargon and relatively little larger context or discussion of how science works.

The structure of undergraduate education often reinforces the transmission of knowledge in the absence of conceptual understanding. Access to higher education is no longer limited to the economically or socially privileged few; more than 55 percent of people over the age of 25 have had at least some college experience (U.S. Census Bureau 2009). Stagnant operating budgets and growing enrollments make large class sizes common at most public institutions. Scientists bear a large responsibility for the teaching of these large science courses. Yet seldom in their academic careers do they receive any kind of pedagogical training that would prepare them for their roles as instructors. Moreover, at many institutions research is more highly rewarded than teaching in promotion and tenure decisions (Anderson and Swazey 1998; Boyer 1991; Shannon et al. 1998). There is little incentive to focus on teaching or innovative pedagogy. Most scientists resort to teaching by the only example they have experienced, the ways in which they themselves were taught as students. Most often this results in didactic lecture courses that emphasize covering content and a prescribed lab practicum that focuses on verifying results. Additionally, scientists are generally assigned to teach the courses that are most closely aligned with their research expertise. The implicit assumption is that expertise within a discipline translates to excellence in teaching topics related to that discipline. This most often does result in presentation of the most up-to-date and cutting-edge research in the discipline. But it less often results in developing scientific ways of thinking or explicitly making interdisciplinary connections that would enable students to see the relevance of the content to their everyday lives. Research in the cognitive sciences and discipline-based education have shown that such teaching methods do little to develop conceptual understanding in most students (DeHaan 2011; Docktor and Mestre 2011; NRC 2000).

Compartmentalized curricula, large class sizes, insufficiently trained instructors with little incentive to improve—is scientific literacy an unobtainable goal for higher education? Even the toughest critics would suggest not. Certainly there needs to be greater attention paid at an administrative level to support restructuring undergraduate curricula such that learning goals are made explicit and assessment of progress toward those goals establishes greater accountability (AAAS 2011; Huba and Freed 2000). Those types of large-scale changes take time and result from concerted efforts of faculty and administrators. But there is mounting evidence that small but powerful changes can be made by individuals or small groups. As scientists, we can start by applying the same level of rigor to our teaching as we do our science. We can more systematically collect evidence of student thinking to measure the effectiveness of our teaching (Handelsman et al. 2004) and revise our methods accordingly. For example, a team of scientists at the University of Washington (Haak et al. 2011) made small changes to the structure of their large-lecture undergraduate biology course over the span of several years and documented the effect on student learning. Beginning in 2002, the course format was gradually changed from a low-structure, lecture-intensive course to a moderate-structure course that made use of clickers and a weekly practice exam, and finally to a high-structure, lecture-free course with reading quizzes, multiple clicker questions, weekly assessments, and active-learning activities.

They compared student achievement across courses, and discovered that all students performed better in the high-structure, lecture-free course design. Further, they noted a disproportionate *increase* in performance by disadvantaged students.

Such results have significant implications for the development of a scientifically minded population. Traditional, lecture-intensive instructional strategies have been shown repeatedly to be less effective than more student-centered pedagogies. And the results above demonstrate that by changing the way we teach, even in large-lecture courses, we can benefit not only all students, but especially those who are the most likely to fail our general science courses. These are powerful results, especially in light of a continually leaky pipeline for women and underrepresented minorities in science and engineering. Knowing that such teaching methods exist and not using them in the undergraduate science courses raises an ethical question. Is it ethically sound to knowingly continue using instructional strategies that are ineffective when there is evidence that other strategies can more equitably benefit a broad population? By using antiquated teaching methods, we are in essence endorsing an exclusionary paradigm of science instruction.

The goals and methods of teaching in science education will need to change if we expect to develop the habits of mind necessary for the next generation to think about and make sense of scientific ideas, claims, and discoveries as they relate not only to their personal lives but also to the surrounding social, economic, and political climate. Within the formal school science curricula in particular, the focus will need to be on demonstrating the *processes* of science in addition to the *products*. The emphasis should be on developing students' understanding of how new scientific knowledge is created, how it fits into broader contexts, and how it interlaces with ethical responsibilities. Scientific literacy is still a noble goal, and we can begin to draw on empirical studies of how people learn science to achieve a scientifically minded citizenry.

The Opportunity Presented by Astrobiology

Philosophers of science, spiritual leaders, and scientists alike have emphasized that astrobiological discoveries, more so than those in any individual science, will have a profound impact on the social, cultural, and spiritual lives of humans. Moreover, many of the advancements associated with these discoveries will be accompanied by significant ethical issues. Yet it has been argued that the current state of the public's general science understanding is insufficient to prepare them for dealing with such profound implications. What, then, should science education look like for future generations? And what role might astrobiology play?

An increasing number of institutions of higher education are now offering astrobiology courses. Offered predominantly at the introductory level, these courses are a welcome rejuvenation to the undergraduate science curriculum. But more than that, they may provide an entry point for changes in undergraduate science instruction. Astrobiology is perhaps the most inherently interdisciplinary discipline in the sciences, and it requires both scientists and students to cross disciplinary boundaries. This has implications not only for how these scientists teach, but also for how students learn in these courses. Moreover, it provides a rich context in which the impact of astrobiology on sociocultural, philosophical, and political aspects of students' daily lives can be fully explored.

As a truly interdisciplinary science, astrobiology requires an instructor to teach outside of her discipline or to recruit other scientists to assist in team teaching the course. If teaching astrobiology as a solo act, the instructor must first make explicit for herself the connections between the disciplines that are necessary to investigate astrobiological questions. She must become familiar not only with the content and vernacular outside her discipline, but also with their theoretical underpinnings and ways of knowing. This requires the instructor to temporarily assume the role of learner, resulting in an awareness of the challenges students might face when drawing connections between concepts. Often this results in instruction that explicitly supports students in integrating their knowledge from the individual disciplines to reason about astrobiological contexts. In the case of team-taught astrobiology courses, instructors must also reach beyond the boundaries of their own disciplines to create a cohesive course. Similar to the solo instructor, the teaching team must work together to identify a framework for teaching the course. They must communicate to one another the key ideas from their individual disciplines required in astrobiology. The result again is a more integrated approach to teaching science at the undergraduate level.

An often-unanticipated benefit of collaborative teaching situations is the increased discussion about the goals and methods of instruction. Most science departments lack a teaching culture in which scientists can exchange their ideas about teaching, pedagogy, and student learning. Undergraduate astrobiology courses provide a unique opportunity to bring together scientists from across departments to discuss and critique one another's approaches to teaching, creating an intellectual space for them to challenge and revise their own ideas about teaching that would otherwise be missing in their everyday academic lives. In the absence of formal pedagogical training, this may be one way to develop instructors' thinking about teaching in a way that will support the development of science literacy in students.

The very nature of astrobiology will require students enrolled in these courses to approach the learning of science from a more interdisciplinary perspective, which is imperative for developing students' abilities to recognize the interrelatedness between the disciplines—a hallmark of scientific literacy. But it will also provide a mechanism for students to examine the impact of science on everyday life. Perhaps in no other discipline are the implications of scientific advances so all-encompassing. Take, for example, the ethical issues associated with excavating a nearby planetary body for natural resources when we have wastefully depleted those on our own planet. Alternatively, consider the economic implications of human exploration missions, or all of the possible implications of the most exciting of all potential astrobiological discoveries—life on another planet. Not only are these provocative issues, sure to capture the interest of a range of students (as they do the working scientists who have contributed to this volume), but they could also realistically occur within a lifetime. Would not a more modern definition of scientific literacy include preparing the next generation to face the questions arising from such advancements?

In short, the goals and methods of science education must evolve to meet the needs of a population that will face not only rapid advancements in science and technology but also complex and deeply personal questions that stem from these advancements. At a minimum, this means that science instruction can no longer depict the scientific enterprise as an objective, static body of knowledge. Rather, we must strive to create opportunities that will equip the next generation with the abilities to access knowledge and use it critically to evaluate the relationship of science to society and everyday life. Astrobiology presents a ready-made instructional context for achieving such goals in the undergraduate curriculum.

Beyond Literacy: Science Preparation for the Next Generation of Scientists

Scientists in all disciplines continue to add to the knowledge in their fields, and in doing so extend the boundaries of those fields, requiring advancement of knowledge at the intersections with other disciplines. For those working at the boundaries, the knowledge and skills required to advance knowledge are different from what were required of their counterparts even 20 years ago. This raises a question about the training of future scientists: Have our methods of teaching become disconnected from the needs of the next generation of scientists?

Scientists have been crossing discipline boundaries for decades. But usually the training preceding their life's work is grounded more heavily in one, perhaps two, scientific disciplines. As astrobiology achieves permanent recognition as its own field, the ways in which future astrobiologists are trained will become of greater concern. These individuals will be charged with identifying a common language, understanding a variety of theoretical and empirical frameworks for scientific investigation, and developing the skills to adeptly and efficiently work across two or more disciplines. Yet the time in which they acquire these skills will necessarily remain unchanged, as degree programs are generally completed within a constrained temporal window. As a result, special attention will need to be paid to the educational and research opportunities afforded to these students as they prepare for their careers as interdisciplinary scientists.

Formal educational experiences for these individuals will need to deviate from the more traditional model for science students. An astrobiologist must draw on understanding from a multitude of disciplines and apply that understanding to complex, integrated problems that often blur the disciplinary lines separating earth, space, and life sciences. It is impractical to believe that a single astrobiologist will simultaneously possess all the knowledge and expertise of a seasoned biologist, geologist, and chemist combined. But she will need a sophisticated understanding of the field, as well as its theoretical underpinnings, to tackle these complex problems. Consequently, foundational coursework for future astrobiologists must, from the very beginning, incorporate context-rich opportunities for students to apply their understanding to interdisciplinary problems. They should be designed in such a way that students gain experience identifying holes in their understanding and learning how to access knowledge needed to solve the problem. Even basic science courses focusing on a single discipline must *explicitly* require these students to draw on and apply knowledge from other areas. Students will no longer be able to sit passively in the classroom as mere recipients of knowledge, but must be actively engaged throughout the learning process.

This will likely require faculty to rearticulate their roles in the classroom. With an emphasis on developing students' abilities to access knowledge and apply it in truly interdisciplinary ways, it will no longer be sufficient to teach through didactic lectures. The backgrounds with which these students enter these courses will be strong, but will be more varied than students in single-discipline programs. Therefore, instructors must gain the skills necessary to gauge the diversity of student understanding prior to instruction—to make student thinking explicit. In addition, instructors will need to grow their views of assessment as a primarily summative activity to consider it also as a formative activity, one that is designed as an integral part of the teaching and learning process to continuously gather data and adapt instructional tasks to develop students' interdisciplinary abilities. Creating rich interdisciplinary opportunities for students to apply their knowledge may also necessitate collaboration among faculty from disparate disciplines to best meet the needs of the students.

Although some programs do integrate education about ethics in science, most often the emphasis is on developing an awareness of ethical research practices. Given the significant implications of discoveries within astrobiology, future scientists in the field will need to be trained not only in ethical research practices but also in communicating with others about the ethical issues surrounding their research. Much research in astrobiology will find itself at the center of public attention, which means that astrobiologists will have the potential to be stewards of fostering scientific literacy in the general public. They should be able to recount their research clearly and invite public discussion of the implications of their work. Although seldom an explicit goal of undergraduate or graduate science programs, excellence in public science communication will be imperative for supporting a modern vision of scientific literacy of the future.

The need for truly interdisciplinary scientists is not limited to astrobiology, but rather will continue to increase at the crossroads of all disciplines as we continue to advance the boundaries of scientific knowledge. As a result, academe will need to continually reexamine and readjust the ways in which the training of future scientists is addressed. The future directions of scientific discovery are more unpredictable than ever before. The next generation of scientists must be equipped not only to adapt to a continually evolving scientific landscape, but also to assist the general public in understanding the implications for everyday life.

Parting Thoughts

Many would argue that a basic understanding of science content is required for an individual to function within a technologically advanced society. Yet consider the number of people who can comfortably operate computers, smart phones, and iPods but have little to no understanding of the science that allows that technology to exist. Just as astrobiology continues to grow as a discipline, so does the number of advances in science and technology. It is already difficult, if not impossible, to identify the minimum knowledge needed to prepare a citizen for a lifetime in a world with expanding scientific boundaries. It is even less clear how to prepare the citizenry for a lifetime where new scientific discoveries directly challenge philosophical and spiritual ideas about the origins of life.

For the first time, science educators will be unable to predict the basic knowledge and skills that will be needed for regular citizens to function in a lifetime of perpetual and exponential scientific advancement, thus necessitating a reexamination of what it means to be scientifically literate. Undoubtedly one aim should continue to be to develop a broad understanding of the nature of science and its relationship to society. However, emphasizing the acquisition of a predetermined minimal level of science content may no longer be possible or realistic. Rather, a more pragmatic scientific literacy for future generations should focus on building the capacity to access scientific knowledge and utilize that information to critically evaluate and solve everyday problems related to science and technology.

References

American Association for the Advancement of Science (AAAS). 1989. *Science for all Americans: A Project 2061 Report on Literacy Goals in Science, Mathematics, and Technology.* Washington, DC: AAAS.

———. 1993. *Benchmarks for Scientific Literacy.* Oxford: Oxford University Press.

———. 2011. *Vision and Change in Undergraduate Biology: A Call to Action.* Washington, DC: AAAS.

Anderson, M. S., and J. P. Swazey. 1998. "Reflections on the Graduate Student Experience: An Overview." *New Directions for Higher Education* 101: 3–13.

Barr, R. B., and J. Tagg. 1995. "From Teaching to Learning: A New Paradigm for Undergraduate Education." *Change* 27 (6): 13–25.

Boyer, E. L. 1991. "Preparing Tomorrow's Professoriate." In *Preparing the Professoriate of Tomorrow to Teach,* edited by J. D. Nyquist, R. D. Abbott, D. H. Wulff, and J. Sprague. Dubuque, IA: Kendall/Hunt.

DeBoer, G. E. 2000. "Scientific Literacy: Another Look at Its Historical and Contemporary Meanings and Its Relationship to Science Education Reform." *Journal of Research in Science Teaching* 37 (6): 582–601.

DeHaan, R. L. 2011. *Education Research in the Biological Sciences: A Nine Decade Review.* Washington, DC: National Academies of Sciences, Board on Science Education.

Docktor, J. L., and J. P. Mestre. 2011. *A Synthesis of Discipline-Based Education Research in Physics*. Washington, DC: National Academies of Sciences, Board on Science Education.

Feinstein, N. 2011. "Salvaging Science Literacy." *Science Education* 95 (1): 168–185.

Haak, D. C., H. R. Lambers, E. Pitre, and S. Freeman. 2011. "Increased Structure and Active Learning Reduce the Achievement Gap in Introductory Biology." *Science* 332: 1213–1216.

Handelsman, J., D. Ebert-May, R. Beichner, et al. 2004. "Scientific Teaching." *Science* 304: 521–522.

Huba, M. E., and E. J. Freed. 2000. *Learner-Centered Assessment on College Campuses: Shifting the Focus from Teaching to Learning*. Needham Heights, MA: Allyn and Bacon.

Hurd, P. D. 1958. Science Literacy: Its Meaning for American Schools. *Educational Leadership* 16: 52.

Khishfe, R., and F. Abd-El-Khalick. 2002. "Influence of Explicit and Reflective versus Implicit Inquiry-Oriented Instruction on Sixth Graders' Views of Nature of Science." *Journal of Research in Science Teaching* 39: 551–578.

Laugksh, R. C. 2000. "Scientific Literacy: A Conceptual Overview." *Science Education* 84: 71–94.

Lederman, N. G. 1986. "Relating Teaching Behavior and Classroom Climate to Changes in Students' Conceptions of the Nature of Science." *Science Education* 70: 3–19.

Liu, S., and C. Tsai. 2008. "Differences in the Scientific Epistemological Views of Undergraduate Students." *International Journal of Science Education* 30 (8): 1055–1073.

Millar, R., and J. Osborne, eds. 1998. *Beyond 2000: Science Education for the Future*. London: King's College London School of Education.

Miller, M., L. Montplaisir, E. Offerdahl, F. Cheng, and G. Ketterling. 2010. "Comparison of Views of the Nature of Science between Natural Science and Nonscience Majors." *CBE Life Science Education* 9 (1): 45–54.

National Research Council (NRC). 1996. *National Science Education Standards*. Washington, DC: National Academy Press.

———. 2000. *How People Learn: Brain, Mind, Experience, and School*. Expanded ed. Washington, D.C.: National Academy Press.

———. 2003a. *Bio 2010: Transforming Undergraduate Education for Future Research Biologists: Committee on Undergraduate Biology Education to Prepare Research Scientists for the 21st Century*. Washington, D.C., National Academy Press.

———. 2003b. *Evaluating and Improving Undergraduate Teaching in Science, Technology, Engineering, and Mathematics*. Washington, DC: National Academy Press.

Parker, L. C., G. H. Krockover, S. Lasher-Trapp, and D. C. Eichinger. 2008. "Ideas about the Nature of Science Held by Undergraduate Atmospheric Science Students." *Bulletin of the American Meteorological Society* 89: 1681–1688.

Roberts, D. A. 2007. "Scientific Literacy/Science Literacy." In *Handbook of Research in Science Education*, edited by S. K. Abell and N. G. Lederman. Mahwah, NJ: Erlbaum.

Roth, M. S. 2012. "Light, Truth and Whatever—Review of 'College: What It Was, Is and Should Be' by Andrew Delbanco." *New York Times*, June 8.

Shamos, M. H. 1995. *The Myth of Scientific Literacy*. New Brunswick, NJ: Rutgers University Press.

Shannon, D. M., D. J. Twale, and M. S. Moore. 1998. "TA Teaching Effectiveness: The Impact of Training and Teaching Experience." *Journal of Higher Education* 69 (4): 440–466.

U.S. Census Bureau. 2009. *2006–2008 American Community Survey 3-Year Estimates*. Accessed February 8, 2012. www.census.gov/hhes/socdemo/education/data/cps/2009/tables.html.

U.S. Department of Education. 1983. *A Nation at Risk: A Report to the Nation and the Secretary of Education*. National Commission on Excellence in Education. Washington, DC: Government Printing Office.

Zeidler, D. L., and N. G. Lederman. 1989. "The Effects of Teachers' Language on Students' Conceptions of the Nature of Science." *Journal of Research in Science Teaching* 26: 771–783.

CHAPTER THIRTEEN

Survival Ethics and Astrobiology

Neville J. Woolf
STEWARD OBSERVATORY, UNIVERSITY OF ARIZONA

Let us start by fantasizing that everything Western science now knows is adequate for understanding the "how" of the Universe. Reality seems so observer-dependent, but the standard Western scientific model is a starting point. There remains the ultimate question, why is there something rather than nothing?, and two associated subordinate questions: why is ours an anthropic Universe? (Barrow et al. 1986), and why does the Universe explore options by a random process?

Biologists have latched onto the second aspect of the subordinate question and its implication that nature's design is developmental rather than an expression of a preexistent form. Cosmologists have latched onto the first aspect, the idea that the Universe started as a single event, the Big Bang, and that an anthropic Universe developed from it. There is an interesting interplay between these two aspects in life. Selection is driven by environment, and environment arises from the Big Bang. Therefore, because the environment selects, this selection has determined the nature of what we find. The randomness that biologists focus on is merely in providing a range from which to select.

Diversity is incomplete here on Earth, because there has not been enough time to explore all biochemical options, but one may well imagine that in perhaps 10^{22} planets, satellites, and other environments where life might develop there is a thorough exploration of possibilities. Thus environmental selection may even explore all possible self-developing life-forms.

Survival and Growth

It has been apparent since *The Origin of Species* (Darwin 1859) that life is an expression of survival. We and all other current life-forms on Earth come from ancestral lines of perhaps 10^{14} organisms, where every one of these survived to reproduce before they expired, though many genes arrived from lateral transfer—from viruses, for example. But the survival that life expresses is not survival of us as individuals, or even of species. Rather it is survival of those things that make survival possible. It is in retention of those things that life's highest priority is expressed in its forms and functions. It therefore becomes, for lack of a better option, the basis of ethics. For those who see existence as a product of God, survival is a value that is directly God-given without distortions from human intermediaries. For those who look for natural values, it is not arbitrary, and it is a shared value.

Exploring this further, we are currently at the top of the greasy pole of life here on Earth. We owe this position to the development and use of manipulative intelligence. When humans understood that something was a potential threat to them, they used their intelligence and manipulative capabilities to kill it. We have done this so well that only one enemy remains; in the words of Walt Kelly, "We shall meet the enemy, and not only may he be ours, he may be us." Through continual obsession with self-interest and lack of foresight and cooperation we pose a serious threat to our own development and even survival.

There is a hierarchy of survival needs that we can follow backward. Survival requires manipulative intelligence. Manipulative intelligence requires learning from experience, and passing it on by teaching. Experience requires time. And to have time, we require survival! The entry point into this loop is the initiation of manipulative intelligence.

How does manipulative intelligence arise? It seems to require mobile life-forms that hone their intelligence by eating or being eaten. Ruth Padel (2009) tells us, "Nature is prodigal with time. She scrutinizes every muscle, vessel, nerve, every instinct, every habit, instinct, shade of constitution. There will be no caprice, no favoring." This stage is the start of learning from experience. Originally it is passed on only by genes. But in higher life-forms it includes parents teaching their offspring. Finally, humans can pass on not only what parents know, but the stored cultural wisdom of the ages in books and digital memory.

The necessary conditions for this stage of learning and teaching are that life has previously arisen with a genetic basis—that is, there is a somewhat mutable code passed to offspring and available for selection. However, life itself can arise only if there is a prior process of growth and crude reproduction available on which a genetic code can be built up. Such growth and reproduction processes are still around, as in hurricanes and forest fires. But those examples have no organizing principles governing their growth, and they are not amenable to adding a code. Presumably in the past there was a transition from a chemical to a biological environment where such a code addition was possible.

The reproduction aspect—even crude pregenetic reproduction like that produced by a forest fire—is crucial because it permits export to new environments. Environments change; those appropriate for growth cease to be appropriate. It is the passage of the reproduced growth form into new environments that allows continuity, even though the former environment no longer supports the process.

Growth processes themselves require an exchange in which their waste material is dispersed so that it does not choke off the acquisition process (in which new ingredients needed for growth are made available from the environment). A prime requirement of continued life is to meet this in perpetuity! This is the essence of sustainability. For an exchange process to occur, there has to have been a prior differentiation of an entity and its environment. That differentiation requires processes of acceptance of that which belongs, defines, and supports, and rejection of that which opposes these processes. These rejections make a coherent "something"—before which there must be something incoherent that can develop coherence. And to initiate the incoherent something, there must be a precipitating event.

Ramps and the Survival Tree

A quite general and abstract version of the transition from nonlife to life is its description as a series of ramps by Dennis et al. (2009). Ramp 1 is initiation, ramp 2 is foundation, ramp 3 is differentiation, ramp 4 is exchange, ramp 5 is growth and reproduction, ramp 6 is information incorporation, ramp 7 is information development, storage, and use, and ramp 8 is integration to ensure the permanence of the exchange process, and transportation to new environments, also to ensure the permanence of the exchange process. (See figure 13.1.)

There is a common confusion that the diversity of life-forms developed because of the random development of genes. Such a view would preclude the possibility of evolution without genes. However, an alternate view is

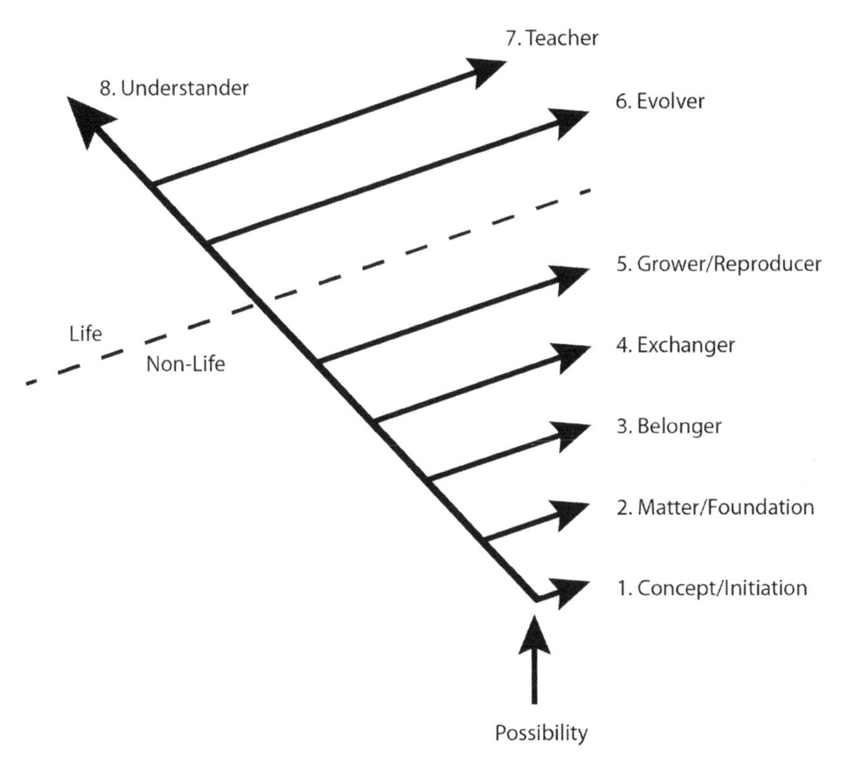

13.1 The Survival Tree

possible. Survival under diverse conditions provides the benefit that drives evolution. This view allows us to follow the survival tree in figure 13.1. Concepts (1) have poor survival ability until they are embedded in a material foundation. Matter (2) in general has poor survival potential unless it has coherence from a relationship of the atoms or molecules to one another (3). But material entities will degrade and diminish unless they can add more to themselves. In general this requires an exchange process (4), in which at least the heat of formation is given away, and often there will be a waste material process. In a cooperative environment, this exchange process will lead to growth, and separate parts of the formation may break off and become independent growth centers (5).

The disadvantage of being unable to modify to work with a changing environment limits pure growers, but if they can develop a mutable genetic system (6), they can change in response to the environment. This is in general adequate for life-forms that do not eat other life-forms, but for those that do, there is a benefit from a nervous system to process data, and for a learning and even a teaching process (7) to benefit from that nervous system. Beyond that, there lies the understanding of the nature of environments and the need to avoid destroying the foundation of the exchange process. This is the highest level of survival: understanding (8).

This set of processes defines a new "great chain of being," with simpler entities that can employ only the lower ramps for survival being at the lower levels of the chain. The role the summit position assigns to us is, as Salk (1992) noted, to use our intelligence "to be good ancestors." That is, we are responsible for the future of our descendants. This is the ultimate basis of our ethics. As St. John of the Cross noted, in our passage to the summit of Mt. Carmel (in his view, the achievement of our divine purpose), there are two errors we can make: to focus on the goods of Earth (we can't take them with us!) or to focus on the goods of Heaven (mind games—we can't take those either) (Kavannaugh and Rodriguez 1991). Similarly, in the Iso Upanishad, produced 3000 years earlier, there is a similar warning about materialism leading to darkness and the focus on mind alone leading to a greater darkness (Swami Nikhilananda 1949). I agree with the need to focus on now.

Ethics and Cooperation

Our lives are not the focus of the process, and our meaning is in what we pass on. All life-forms are renewable resources, and the evolutionary process ensures that the new model will have some defects removed. Every part of existence is an expression of the whole (holy), but humans are not individually so different. The destruction of consciousness, and in particular the killing of humans after their mental processes have developed, is to be avoided where practical because of its brutalizing effect on the community, not because human life is any more sacred than any other part of existence.

If we examine the various ways in which we currently fail to be good ancestors, the one that stands out most is our prodigiously large and unnecessary population, its associated over-rapid use of resources and equally over-rapid production of waste products at a faster rate than the environment can transform them. We could have all the people we want, but we cannot afford them all at the same time! By that large population we have become like locusts upon the Earth, consuming everything in our path without any care for our own future.

Some question whether our very nature will make the necessary cooperation impossible. In the past, progress has been hindered by groups trying to hold on to power and social cultural structures. That makes cooperation fail. We will have to understand this, and eliminate it in future decision making. I believe that if we cannot do that, humanity is doomed. Lest there be any doubt, I find all cultures to be inadequate solutions to human problems. When I look at their preparation for the future, their care of children, women, homosexuals, those who are victims of crime and those who commit crimes, I see no adequacy anywhere. The disparity in justice between the rich and the powerful and the poor and the helpless makes for a devaluation of law and law enforcement, and for class conflict.

Individual groups have in the past, and unfortunately in the present too, deliberately tried to maintain irrelevant cultural and religious forms by keeping up an unnecessary production of humans. I do not recommend us killing one another. This combined with famine and disease has produced sufficient death to be life's past answer to overpopulation. Rather, if we care for each other so that we all have no need for individual families to care for us, we can all afford to let the population shrink. But reduction must be at such a rate that we do not overburden the smaller future population in the process. Cultures and religious forms all die in their time and are replaced by forms appropriate to the new social needs and understanding. We do need to hold on to experience. That is, the processes of continued research, development, information storage, and teaching are essential ingredients of a values system.

Let us dispose of the idea that the population explosion can be solved by sending people into space. A generation lasts about 20 years and in that time could easily produce both a replenishment of Earth's population and an equal number needing to be exported into space. With about 7 billion people on Earth, we would therefore need to export 350 million people per year into space, or approximately 1 million per day. That is, every day we would have to gather together a population the current size of greater Tucson, Arizona, and ship them off. If we were to make rockets that held 12 people each, we would be sending up a rocket every second, day and night, forever. It is an absurd notion. So until we have put our own house in order, it seems inappropriate to expand elsewhere, and so create other groups with this same unsolved problem. Fix it once, here, and it remains fixed.

In the same vein, Earth's resources are enough for a very long time, provided we use them at a reasonable rate. It does not make sense to put energy and expense into importing resources from elsewhere, when that energy might be better spent in dealing with the problems on our home planet. The same logic may well apply to other civilizations in space, which have to operate under a constraint of finite resources.

Values and Cooperation

Rousseau (1762) pointed out that might does not make right. If a thief demands my wallet, I have no moral responsibility to give it him. So then what does make right? Here it is suggested that long-term survival of understanding is the appropriate determiner. Others seek for rigid rules, which they usually find by selective reading from religious literature, focusing on that which they want, finding the appropriate words to justify themselves, and ignoring that which is in opposition to what they want to do. I believe that good religious literature is always self-contradictory precisely because its goal must be to place responsibility on the doer. But the self-righteous hold an illusion that they are religious.

We must reject the application of rigid rules because it has the unfortunate result that, regardless of how well intentioned the writers of the rules were, when the consequences are now seen to be detrimental, the step of reversing the action is forbidden! Rule following fails, because it is only self-correcting if we can rewrite the rules. Some who favor rigid rules have scoffed at the notion of inferred consequences being an appropriate determiner of actions. Yes, good intentions do not automatically produce good results, and lack of concern whether one's actions will harm others may not automatically produce bad results. However, because one has no good alternative to choosing one or the other, the preference is obvious. If one monitors what is happening, then there is a good likelihood of correcting a misstep.

When Peirce and James developed pragmatism at the end of the nineteenth century, there was an outcry from people who did not understand that long-term effects are an important aspect of pragmatic values. If we are looking for pragmatic ways of doing things that will make us a good ancestral group, we will not do things that are socially destructive. We must depart from the path of unhindered activity regardless of its social consequences, to one where the inferred consequences are determiners of the permissible paths (see, in this volume, Bedau, Stoeger, and Sullivan).

We all would like the ability to make choices and to act without consultation. Unfortunately that freedom is all too divisible. If we take the land surface of the Earth and divide it between all the people, we each have a patch 160 yards on a side. About 20 percent of this is desert, and large other percentages are mountains, swampland, tundra, ice caps, and jungle. All the vegetable and animal matter we eat has to be produced on it. Some is needed for roads, hospitals, shops, parks, and schools. Rather little is good arable land. Further, we share it not only with all humanity, but with all of nature too. These needs limit our self-expression. Reducing Earth's human population by a factor of 100 or so would make us all less limiting to each other. We are forced to joint action and joint decision making because we are so interdependent in the use of land.

A particular aspect of decision making is that humans are substantially motivated by what happens to them. Unless they are given a stake in the success of the enterprise in which they are engaged, they will lack the motivation to help the enterprise improve. There is a huge difference between collaboration and cooperation. In collaboration we merely work together. The task and activities are externally determined. In cooperation, the operation, the production and nature of the opus, is jointly developed and determined. Everyone has a stake in it.

The ramps of development provide a model for cooperative organization, with eight units that for short-term activities are semiautonomous. There must indeed be creators, dreamers, and initiators that play a ramp 1 role. There is also a need for a ramp 2 unit that operates facilities and maintenance. Operations, the ramp 3 section where the product arises, needs its own unit. The exchange process of ramp 4 is concerned with purchases and sales, or their social exchange equivalents. It is the financial unit. Transportation is a key factor in production, purchases, and sales, but it is also needed in handling growth and in the initiation of new units of the organization. It follows ramp 5 roles. Ramp 6 is information technology. It handles the way the organization is able to store information and retrieve it with speed and ease. Ramp 7 is where the research and development are carried out that direct the future paths. It must also be responsible for teaching all in the organization about what has been learned, and how it may affect what the organization does in the future. Finally, there must be a planning unit for ramp 8 that is concerned with the long-term future of the organization.

These units fall into two groupings. Ramps 2, 3, 5, 4, and 8 must work together to handle immediate tactical needs. Ramps 1, 6, 7, 4, and 8 must work together to handle strategic goals and the future. Ramp 1 activities need to be fed into all other parts of the organization, to learn of issues and problem areas and to explore alternate ways of acting.

Robert Fulghum (1990) pointed out that cooperative habits need to be started very early. The practice of collaboration and learning its benefits needs to start in the home and at kindergarten. It is possible to start children on this path in elementary school by giving them a substantial responsibility for the social aspects of their schooling, with activities associated with each ramp. An experiment on these lines has been in progress in Murano (near Venice), Italy, for six years, and is in process of exportation to other countries.

Equally, a few companies have been remodeled to work on the same pattern. I see a future in which all human activities are handled in this way. By our lack of necessary and coordinated activities, we pose threats to our own optimal development and even survival. To achieve long-term survival, that coordination will require all humanity to have a stake in the enterprise, and a feeling that their needs are appreciated, that they will be supported and their voices heard. Rewards in the form of resources, and their absence, will reinforce the process. In return all people will be expected to support their social structure. That requires a new start, a new set of values, and activities to match. It is all too easy to have a revolution that merely replaces people and apparent forms without changing the underlying structure, values, and behaviors.

Conclusion

I have argued that our very nature as a survival form forces us to have values that perpetuate survival. These values derive from a sequence of developments that have been the basis for our existence. The connection to astrobiology comes from the speculation that other advanced biological entities share these developments. The crucial ones for us to focus our attention on start with ramp 3, where there is a requirement for sharing this survival goal and its implications. For ramp 4, we need environments that are energized to permit exchange. The two implications are for conservation in each environment so that exchange is possible, and propagation to new environments to allow for environmental fluctuation with time. If we take care of these needs, most issues of ramp 5 (growth) and ramp 6 (information incorporation) will take care of themselves. We are left with a need for the ramp 7 activities of research and information storage and access, so that learning and teaching its results can take place. Finally, for ramp 8 we have to plan so that life is integrated with its environments, and is spread in such a way that changing environments always still permit some life to continue the process of learning and developing.

Survival values developed from this astrobiological view are based on using the best available information to predict both long-term and short-term consequences. The relationship to space activities arises mainly from the idea that we need to get our own house in order rather than export problems. Learning as a principal space activity minimizes problems. Competitive acquisition of space resources maximizes our exporting of our problems. In restricting what we do to learning, our terrestrial problems need resolution only once.

Acknowledgments

Many of these ideas have been honed in e-mail exchanges and rare time together with my collaborators, Lynn Claire Dennis, Robert Gray, Louis Kauffman, and Jytte Brender McNair. I thank them for this. I would also like to thank the NASA Astrobiology Institute for the support of my work in astrobiology, while in no way implying that it or NASA share any of the opinions expressed here.

References

Barrow, J., F. Tipler, and J. A. Wheeler. 1986. *The Anthropic Universe*. New York: Oxford University Press.

Darwin C. 1859. *Origin of Species*. London: John Murray.

Dennis, L., R. Gray, L. Kauffman, J. B. McNair, and N. Woolf. 2009. "A Framework Linking Non-Living and Living Systems: Classification of Persistence, Survival and Evolution Transitions." *Foundations of Science Journal* 14: 217–238.

Fulghum R. 1990. *All I Really Need to Know I Learned in Kindergarten*. New York: Villard Books.

Kavanaugh, K., and O. Rodriguez. 1991. *St. John of the Cross: Collected Works*. Washington, DC: Institute of Carmelite Studies.

Kelly, W. 1952–1953. *An Introduction to Pogo Papers*. New York: Simon and Schuster.

Nikhilananda, S. 1949. *Upanishads*. Vol. 1. New York: Harper.

Padel, R. 2009. *Darwin: A Life in Poems*. New York: Knopf.

Rousseau, J. J. 1762. *Social Contract*. Amsterdam: Chez M. M. Re.

Salk, J. 1992. "Are We Being Good Ancestors?" *World Affairs* 1 (2): 16–18.

APPENDIX

Astrobiological Risk

A Dialogue

Steven A. Benner
FOUNDATION FOR APPLIED MOLECULAR EVOLUTION

Neville J. Woolf
STEWARD OBSERVATORY, UNIVERSITY OF ARIZONA

The educated population has a considerable wariness about astrobiology. If we discover life elsewhere and bring it back to Earth, is there not a possibility that it will cause a catastrophic pandemic? If an advanced life-form comes to Earth because it has detected our presence, is there not a risk that it will enslave us or eat us? And if we create a microbial life-form in the laboratory, or even a highly interactive chemical, might that not present a similar risk? This Appendix to the volume expands on the chapters in this book, which present the thoughts of individual scientists and philosophers engaged in the consideration of the societal implications of astrobiology through the lenses of their lifelong research. In this Appendix, two astrobiologists focus on a consideration of astrobiological risks and engage in an informal dialogue about them, which grows out of their personal experience in different areas of research.

As the contributions to this volume demonstrate, astrobiology includes participants trained in various disciplines, and the authors of this dialogue have diverse backgrounds. Steven Benner is a biological chemist. His interest is in understanding life through chemistry, joining traditions in natural history to traditions in the physical sciences. He has developed base pairs that can substitute for the base pairs AT and CG in DNA; these are used each year to personalize the care of 400,000 patients suffering from AIDS or hepatitis. He has developed ways of predicting protein folding, resurrected ancient proteins, and explored the evolution of ruminants and the temperature regime at which early terrestrial life existed. In his book *Life, the Universe and the Scientific Method* (2009) he discusses these issues in a form appropriate for both students and a popular audience. Some of the ideas expressed here are developed further in that book.

Nick Woolf was trained as a physicist, became an astrophysicist, and through curious circumstances came to understand biology. When he wanted to place telescopes on Mt. Graham in Arizona, he was required to understand the issues of the survival of a red squirrel subspecies. From a book by G. G. Simpson (1983) he was directed to biogeography (MacArthur and Wilson 1967), and he taught himself population dynamics modeling to understand how population dynamics respond to environmental changes and the presence of a sympatric species. When an opportunity came to design space telescopes to look for Earth-like planets and signs of life, he started exploring how the products of metabolism could be used in the search. Thus, he started to try to understand what life is and what its signatures are.

Benner and Woolf first interacted in the development of the second NASA Astrobiology Roadmap (Des Marais et al. 2003). Later they spent time (separately) on the National Academy of Sciences Committee on the Origin and Evolution of Life. They agree that the most significant aspect of science is prediction, but that one must accept the nature of prediction in the forms found in nature. After all, some of the most important answers for humanity are just yes or no!

Dialogue

NW: As I understand your description of what biological chemists do, they first analyze the way the atoms in a particular biological molecule are put together, and then how that molecule interacts with others in life. Then they synthesize the biomolecule, putting each atom in its appropriate place, held together with other atoms in the right chemical bonding. Synthesis of the biomolecules allows them to be sure that their understanding is correct. Finally they synthesize a variant, so as to explore how its properties differ from the original compound, and to see whether that too confirms their original understanding. How risky is biochemical synthesis for humanity?

SB: Let me take this stepwise. First, scientists can apply the process of analysis, taking a system apart and making a list of its parts. When reductionism goes to the atomic level, this path is well trodden; many biological chemists have done this type of thing. Deciding how the parts interact in a living system is an entirely different matter. We still have no general way to understand the "how."

As for synthesis, reassembling the atoms to get a piece of a natural biomolecule from a living system is how we know that we have got right the reductionist analysis. If we can reconstitute the system and the reconstituted system works like the original, we know that we have not missed any parts, and our proposed model for the arrangement of the parts in the original is correct.

As for routes to understanding how the parts of the living system work in the whole, this is not a solved problem. We do not, for example, understand how perturbation in the structure of any biomolecule will change the fitness of a host organism. On the other hand, we are a bit better in predicting how the perturbation in the structure of a biomolecule might change its molecular behavior—for example, its solubility in water, or its interaction with other molecules dissolved in water.

To develop an understanding in this difficult frontier of biological chemistry, we make an *unnatural* compound that has a *different* arrangement of atoms but that, we hope, will reproduce the behavior of the natural biomolecules. We might even reinsert the synthesized molecule into a living organism and see if the outcome is predicted from our model. That shows whatever part of the original understanding is correct, but at a very fundamental level. If a designed molecule works, even if it is different from the natural molecule, then we know that the theory behind the design that relates chemistry and physiology is correct.

Finally there is the issue of usefulness. Our usefulness is not in increasing the number of humans alive, but in helping and preparing humanity to move to a new situation in which we can focus on the quality of life, not the quantity at any one time. For example, it would be nice to not spend one's old age going from doctor to doctor trying to understand why you do not feel well. A crisp diagnostic process would come from a better understanding of the relation between your chemistry and your physiology.

NW: But there is still the issue of the risk while you are learning this.

SB: There are individual risks and systemic risks. Of course, any experiment carries individual risk. A patient taking a drug incurs a risk of a side reaction. An experiment done in the laboratory might lead to the release of a toxic metal into the environment. At a more trivial level, you might die in an automobile accident driving to the lab to do the experiment.

What concerns most people who hear the terms "biotechnology," "artificial life," or "synthetic biology" is that what will emerge is a life-form that will cause human disease. Especially bad in this time of "swine" H1N1 flu is the image of a synthetic pathogen escaping from the laboratory and causing a pandemic.

Balancing these is the risk of *not* doing such research, and therefore *not* learning more about the chemistry behind biology. A historical example illustrates that risk. In 1975 the city of Cambridge, Massachusetts, banned recombinant DNA technology within its borders to manage what it saw as a danger in cloning genes from one organism to another. We now know that their concern was overwrought. But the historical lesson is clear. If that ban had been worldwide, in 1985 we would not have had the technology to understand AIDS, the nature of the virus that causes it, or the drugs that we now have to manage it. The city's attempt to manage the known minor risk would have aggravated an unknown (at the time) risk that was much larger.

Today, we are using synthetic biology as the next level of technology past biotechnology to develop the understanding that may be needed to manage the next risk. Remember, only 90 years ago there was an influenza outbreak that caused 20 to 40 million deaths, more deaths than caused directly by World War I.

This process, learning more and more that allows us to manage future challenges better and better, is distinctly human. Our process of learning and passing on the learning to our descendants is a kind of transmission of acquired characteristics. This is our nature. Human evolution is no longer driven by mutation. Learning is our evolutionary path, and it includes some risks inherent in learning. We should use our intelligence to minimize the risk, but we have left the path of ignorance behind. Any hazard must be juxtaposed against the potential benefits that come from learning.

NW: Let's break this issue into three pieces. Some are concerned about the risks of importing a life-form from elsewhere. Some are concerned about the risks of creating new life-forms here on Earth, and I have concerns about whether even the manufacture of new biochemical compounds has some hidden risks. Why don't you lay out a way for exploring these issues.

SB: Let us take the second concern first, and consider creating new forms of life here on Earth. Of course, this happens every time we give birth to a child, which involves a rearrangement of genes in a population in a new context. Biotechnology allowed us to move genes more easily between species, and move them by design. This has created the new field of bioengineering, and the engineering component of the field of synthetic biology. Still more dramatic is the construction of a life-form from atoms, not drawing at all upon anything that natural biology has given us.

A Venn diagram can be used to explore the relative hazards of these activities. Activities within the lower circle use standard terran biochemistry, developed over the past four billion years. The farther out of the circle, the less similar the biochemistry. The upper-right circle represents organisms capable of evolving. Those outside cannot evolve, and represent no more hazard than a toxic chemical. The upper-left circle represents those systems that are capable of being self-sustaining. They live without human intervention. Those outside that circle survive only if fed by humans. They represent no more hazard than a pathogen that will die once released from the laboratory. We are at greatest risk from organisms that, like terrestrial organisms, lie within all three circles.

NW: Following on from this, it is interesting that astronomical environments within the Solar System are so different from our own. For example, the free oxygen in Earth's atmosphere is not found elsewhere, the ultravio-

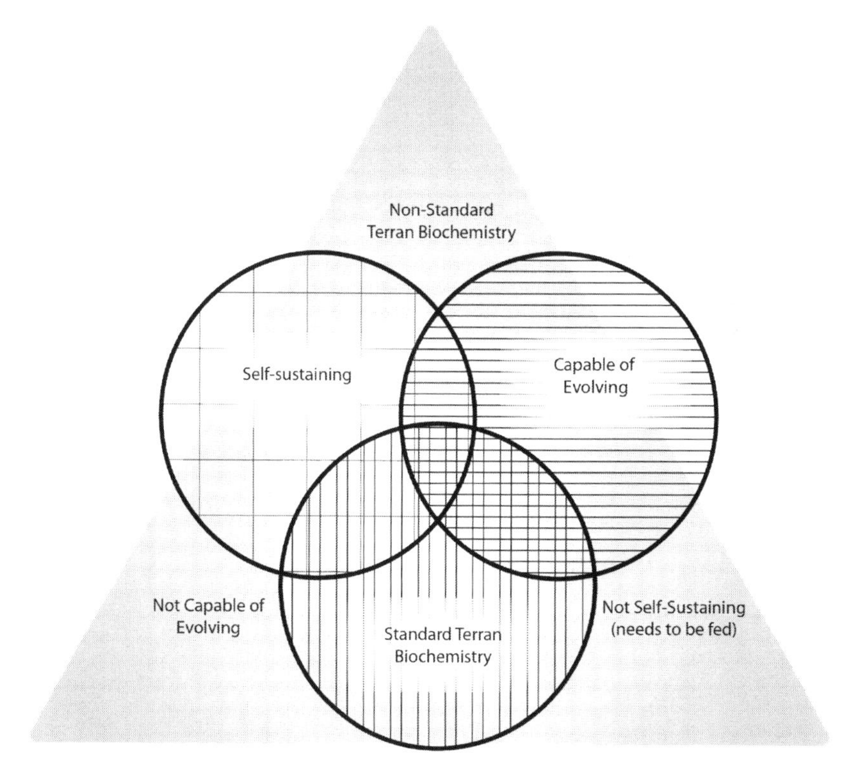

App.1 Astrobiological Risk

let radiation from the Sun is mitigated here in a way not found on other bodies, the risk of breaking apart in liquid water is greater in Earth's ample liquid water, and Earth's temperature regime is wide, which includes conditions that could be either too hot or too cold for alien forms of life.

These differences all represent difficulties that Earth's environment poses to organisms from elsewhere. Even advanced life-forms would likely have difficulty in living here without a spacesuit.

Further, many of our fears of aliens with advanced technology represent concerns about other humans, rather than appropriate expectations for an advanced life-form that might make it here and has undergone necessary behavioral changes to survive technological development. We and other terrestrial life-forms do not make good slaves—we do not even make good servants! Robotic forms are better at carrying out orders. And Earth is a poor environment for obtaining anything, because the gravitational well we are in makes anything taken elsewhere very expensive in terms of energy consumption.

SB: Evolution has shaped all terrestrial organisms to survive here. Extraterrestrial organisms have difficulty not only with survival. They also must metabolize, repair their structure using environmental materials, and reproduce. Photosynthesis at the Earth's surface produces various carbohydrates, which in turn can give back energy by combining with oxygen. If an organism cannot either do that, or photosynthesize from available materials (such as the tiny amount of CO_2 in our atmosphere), it has to make do with a much smaller amount of energy available from metabolism from existing organic molecules. And even metabolism is an evolved skill.

NW: This is rather like the issue of whether a terrestrial organism can colonize a new environment. Most of them will fail. But if one survives, it will find an environment that may well lack its natural enemies and it is likely to thrive—like rabbits in Australia, the mongoose in Hawaii, and the AIDS virus in humans.

SB: We do not fully understand the hazard to life-forms that arise from introducing new life-forms from outside. In general, placing an organism into an environment for which it has not adapted leads to its extinction. The rabbits that you mentioned survived because Australia is not all that different from the environment from Europe. The atmosphere is similar, the amino acids in the food that animals eat are the same, and so on. Further, the European predators of the rabbits are not generally brought along. Had Australia been a truly alien environment, rabbits would not have survived. For the same reason, a life-form evolved on Vulcan (a fictional planet in *Star Trek*) would probably not find the food available on

Earth very nourishing, because the molecules that nourish Vulcans have not evolved on Earth.

NW: In your book you point out how even the simple need to get nutrition from cellulose required animals to develop a genetic change in RNAse, and to start chewing cud 40 million years ago. A similar problem occurred when trees first became common 360 million years ago, and had lignin in their cell walls. There was no way for organisms to digest lignin, and so when trees died they were buried rather than digested (giving rise to coal), and the oxygen content of the atmosphere almost doubled as a result. Even now lignin requires fungi to digest it. So, you seem to be suggesting that because the detailed chemical bonding of life's molecules is likely to be idiosyncratic to the environment, we have more to fear from organisms developed on Earth than from those coming from elsewhere.

SB: The greatest chance for hazard comes from a system that is self-sustaining here, uses standard biochemistry, and is capable of evolving. This is of course the goal that Craig Venter and Hamilton Smith (see Glass et al. 2006) are pursuing in their synthetic biology research, which has the goal of producing an artificial cell by rearranging genes from natural biology. If the Venter-Smith minimal cell were to develop from its presently nonpathogenic form to a pathogenic form, it would present the same hazards to us as a pathogenic organism that has developed naturally from a nonpathogenic microbe to feed on us. Those hazards, though not absent, are not large compared to those presented by many natural nonpathogens that cohabit the Earth with us.

NW: So let us discuss the risk of making something that is not a life-form, something that might seem very ordinary, but that might interact with human biochemistry.

SB: The question is, can a nonoptimized system prepared by a chemist compete with an optimized system evolved after four billion years? In principle, if you swallow an alien organic compound, it may kill you. Simple compounds such as cyanide can. Complex compounds such as strychnine can. Every now and then, a small compound pops up that has devastating impact on an individual who takes it. A recent example emerged with people who were attempting to create new illicit "designer drugs." One was stumbled upon that created acute Parkinson's disease. Anyone who took it was paralyzed, conscious but unable to move (Davis et al. 1979). But this is rare.

NW: I accept that a simple alien molecule will not do that. But you do experiments on protein folding, and there is at least a suspicion that a different folding of a protein is somehow involved in the various brain diseases, such as "mad cow" disease. There must be some risks of newly created proteins somehow interacting with living cells and resulting in the production of something harmful.

SB: Yes, for individual risk (as opposed to systemic risk), this is always a possibility. One might create in a laboratory a protein that, if you eat it, causes you to develop mad cow disease. We know that there are not many proteins that do this. And your disease would not be transferable (except to cannibals). But the risk is there.

NW: So there is a need for some kind of protocol to protect us from something unexpectedly nasty coming out of your kind of lab. I say that in contrast to other labs that are interested in the possibility of generating deliberate nastiness toward humans. What precautions do you take?

SB: In 1975, guidelines were initially laid out to manage individual and systemic risk that comes from biotechnology generally. These have been revised as technology advances. Everyone in my organization is trained in these guidelines, and we apply them rigorously.

Conclusion

Life is not a finished process. Our own path of development is by learning and passing on what we have learned. That path has carried risk since the first human started to control fire almost a million years ago. Yet this path of understanding has been very successful. There are now more humans living lives of comfort than there were humans altogether in the early years of this development.

We are not saying that the development is risk-free. Fortunately, Earth has been continuously collecting materials from Solar System environments for the past four billion years. They have arrived as meteorites and as interplanetary dust particles. Through these, life has already been exposed to much of what might be potentially harmful from Earth's neighborhood.

There is a need for protocols for what we send elsewhere, and what we let loose from the laboratory here. The dangers are worse from unstable people deliberately trying to cause harm. Yet when we look at the five deaths caused by the deliberate multiple exposure of the public to anthrax (Centers for Disease Control 2002), and compare it to the 168 deaths from a single deliberate explosion of ammonium nitrate and fuel oil in Oklahoma City, it would seem that there is more to fear from the unstable having access to explosives and guns.

At a different level, the pressure to generate public interest in scientific research in order to generate research funds is itself a problem. If the funds are just enough for the experiment, then for the headstrong experimenter there is none left over for risk abatement. The most likely first consequence of this will be that some scientists will accidentally kill themselves! We will be forewarned. As Alexander Pope wrote, "A little learning is a dangerous thing; drink deep, or taste not the Pierian spring: there shallow draughts intoxicate the brain, and drinking largely sobers us again."

References

Benner, S. A. 2009. *Life, the Universe and the Scientific Method.* Gainesville, FL: FAME Press.

Centers for Disease Control. 2002. "Bioterrorism-Related Anthrax: International Response by the Centers for Disease Control and Prevention." *Emerging Infectious Diseases* 8 (10): 1056–1059.

Davis, G. C., A. C. Williams, S. P. Markey, et al. 1979. "Chronic Parkinsonism Secondary to Intravenous Injection of Meperidine Analogues." *Psychiatry Research* 1 (3): 249–254.

Des Marais, D., et al. 2003. "The NASA Astrobiology Roadmap." *Astrobiology* 3: 218.

Glass, J. I., N. Assad-Garcia, N. Alperovich, et al. 2006. "Essential Genes of a Minimal Bacterium." *Proceedings of the National Academy of Sciences* 103 (2): 425–430.

MacArthur, R. H., and E. O. Wilson. 1967. *The Theory of Island Biogeography.* Princeton, NJ: Princeton University Press,

Simpson, G. G. 1983. *Fossils and the History of Life.* New York: Scientific American Books / W. H. Freeman.

Further Reading

Bedau, Mark A., and Paul Humpheys, eds. *Emergence: Contemporary Readings in Philosophy and Science.* Cambridge, MA: MIT Press, 2008. 464 pp.

Bedau, Mark A., and Emily Park, eds. *The Ethics of Protocells: Moral and Social Implications of Creating Life in the Laboratory (Basic Bioethics).* Cambridge, MA: MIT Press, 2009. 392 pp.

Beech, Martin. *Terraforming: The Creating of Habitable Worlds.* Berlin: Springer, 2009. 292 pp.

Benner, S. A. *Life, the Universe and the Scientific Method.* Gainesville, FL: FAME Press, 2009. 312 pp.

Bertka, Connie Roth, ed. *Exploring the Origin, Extent, and Future of Life: Philosophical, Ethical and Theological Perspectives.* Cambridge: Cambridge University Press, 2009. 336 pp.

Brown, Neville. *Engaging the Cosmos: Astronomy, Philosophy, and Faith.* Brighton, UK: Sussex Academic Press, 2006. 367 pp.

Chela-Flores, Julian. *A Second Genesis: Stepping-Stones towards the Intelligibility of Nature.* Hackensack, NJ: World Scientific Publishing, 2009. 248 pp.

Davies, Paul. *The Goldilocks Enigma: Why Is the Universe Just Right for Life?* London: Penguin Press, 2006. 336 pp.

Dick, Steven J. *The Biological Universe: The Twentieth Century Extraterrestrial Life Debate and the Limits of Science.* Cambridge: Cambridge University Press, 1999. 600 pp.

Gale, Joseph. *Astrobiology of Earth: The Emergence, Evolution, and Future of Life on a Planet in Turmoil.* Oxford: Oxford University Press, 2009. 240pp.

Grinspoon, David. *Lonely Planets: The Natural Philosophy of Alien Life.* New York: HarperCollins, 2004. 480 pp.

Hargrove, Eugene C. *Beyond Spaceship Earth: Environmental Ethics and the Solar System.* San Francisco: Sierra Club Books, 1986. 336 pp.

Harris, Philip. *Space Enterprise: Living and Working Offworld in the 21st Century.* Berlin: Springer-Praxis, 2008. 620 pp.

Impey, Chris, ed. *Conversations about Astrobiology.* Cambridge: Cambridge University Press, 2010. 418 pp.

———. *The Living Cosmos: Our Search for Life in the Universe.* New York: Random House, 2007. 393 pp.

Impey, Chris, and Catherine Petry, eds. *Science and Theology: Ruminations on the Cosmos*. Notre Dame IN: University of Notre Dame Press, 2004. 184 pp.
Jakosky, Bruce. *Science, Society, and the Search for Life in the Universe*. Tucson: University of Arizona Press, 2006. 160 pp.
Lunine, Jonathan I. *Astrobiology: A Multi-disciplinary Approach*. Old Tappan, NJ: Benjamin Cummings, 2004. 450 pp.
Matloff, Gregory, Les Johnson, and C. Bangs. *Living Off the Land in Space: Green Roads to the Cosmos*. New York: Copernicus Books, 2007. 250 pp.
NASA Astrobiology Roadmap. http://astrobiology.nasa.gov/roadmap. 2008. 26 pp.
Sullivan, Woodruff T., III. *Planets and Life: The Emerging Science of Astrobiology*. Cambridge: Cambridge University Press, 2007. 626 pp.

Contributors

Mark A. Bedau, Reed College. Dr. Bedau received his PhD in philosophy from the University of California, Berkeley. He studies complex adaptive systems and their philosophical and scientific implications for emergence and open-ended evolution of life, mind, technology, and synthetic biology. He also studies their social and ethical implications, with special focus on creating new forms of life in the laboratory, including synthetic cells and protocells. He is a pioneer in empirically measuring the creativity of natural and computational evolving systems, and in evolving complex biochemical systems to have desired useful properties. For over a decade he has been editor in chief of the journal *Artificial Life*, and he regularly teaches in the program in Foundations and Ethical Implications of the Life Sciences at the European School of Molecular Medicine in Milan, Italy. He has co-authored a number of books, including *Emergence* (2008, MIT Press, with Paul Humphreys), *The Nature of Life* (2010, Cambridge University Press, with Carol Cleland), *Protocells: Bridging Nonliving and Living Matter* (2009, MIT Press, with S. Rasmussen, L. Chen, D. Deamer, D. C. Krakauer, N. H. Packard, and P. F. Stadler), and *The Ethics of Protocells: Moral and Social Implications of Creating Life in the Laboratory* (2009, MIT Press, with Emily Parke).

Steven A. Benner, Foundation for Applied Molecular Evolution (FAME). Dr. Benner has been in the forefront of redefining science fields. In 1991 he helped found evolutionary bioinformatics. He established paleomolecular biology to resurrect ancestral proteins from extinct organisms for study in the laboratory. He invented dynamic combinatorial chemistry to generate strategies to discover molecule therapeutic leads. He initiated synthetic biology as a field, and his research led to the synthesis of artificial proteins to support the first artificial chemical system capable of Darwinian evolution. In 2004 Dr. Benner founded FAME. Dr. Benner has won numerous awards. He received his BS and MS in Molecular Biophysics and Biochemistry from Yale in 1976 and his PhD in chemistry from Harvard in 1979.

Carol E. Cleland, University of Colorado (CU) Boulder. Dr. Cleland is a professor of philosophy at the University of Colorado. She arrived at CU Boulder in 1986, after having spent a year on a postdoctoral fellowship at Stanford University's Center for the Study of Language and Information. She received her PhD in philosophy from Brown University in 1981 and her BA in mathematics from the University of California

(Santa Barbara) in 1973. From 1998 to 2008 she was a member of NASA's Institute for Astrobiology. Professor Cleland specializes in philosophy of science, philosophy of logic, and metaphysics. Her current research interests are in the areas of scientific methodology, historical science, biology (especially microbiology, origins of life, the nature of life, and astrobiology), and the theory of computation. Her published work has appeared in leading journals in philosophy and science. She is co-editor (with Mark Bedau) of *The Nature of Life: Classical and Contemporary Perspectives from Philosophy and Science* (forthcoming from Cambridge University Press) and is currently finishing a book (*The Quest for a Universal Theory of Life: Searching for Life as We Don't Know It*), which is under contract with Cambridge University Press. Professor Cleland's website is http://spot.colorado.edu/%7Ecleland/index.html.

Martinez J. Hewlett, Professor Emeritus, Molecular and Cellular Biology, University of Arizona. Dr. Hewlett earned a bachelor's in chemistry at the University of Southern California in 1964. After working for five years as a research biochemist at the VA Hospital in Sepulveda, California, he returned to academia, earning his PhD in biochemistry at the University of Arizona in 1973. Dr. Hewlett was a postdoctoral fellow in the laboratory of David Baltimore at Massachusetts Institute of Technology, where he specialized in the molecular biology of first poliovirus and then the emerging family of bunyaviruses. He joined the faculty of the University of Arizona in 1976, becoming an emeritus professor in 2003. Since closing his research laboratory, Dr. Hewlett has concentrated his efforts on teaching, creative writing, and research in the philosophy of science, particularly the spiritual aspects of existence. He has published one novel (*Sangre de Cristo: A Novel of Science and Faith*) and is working on a second. He is a lay member of the Dominicans and a member of LENS, a diverse group of futurists. Now living in Taos, New Mexico, he writes, consults, and does some teaching.

Nishant Alphonse Irudayadason, Jnana-Deepa Vidyapeeth. Dr. Irudayadason is a professor of philosophy at Jnana-Deepa Vidyapeeth: Pontifical Institute of Philosophy and Religions (JDV), Pune, India. He completed his doctoral studies on postmodernity and ethics at Université Paris-Est, France. His regular courses include Research Methodology, History of Contemporary Philosophy, Ethics, Politics and Hermeneutics. He has written two books and 18 articles and attended numerous national and international conferences. He is a regular contributor to contemporary political analysis in *Light of Truth*, a bimonthly published in Kochi, Kerala, India. His areas of interest include ethics, postmodernity, contemporary western and Indian philosophies. He is also the director of the JDV Centre for Ethics.

Jonathan I. Lunine, Cornell University. Dr. Lunine is director of the Center for Radiophysics and Space Research, and David C. Duncan Professor in the Physical Sciences at Cornell. He is the David Baltimore Distinguished Visiting Scientist at NASA's Jet Propulsion Laboratory. His research interests center broadly on the formation and evolution of planets and planetary systems, the nature of organics in the outer Solar System, and the processes that lead to the formation of habitable worlds. He is an interdisciplinary scientist on the Cassini mission to Saturn and on the James Webb Space Telescope, as well as co-investigator on the Juno mission, which is en route to Jupiter. Dr. Lunine is the author of over 200 scientific papers and of the books

Earth: Evolution of a Habitable World and *Astrobiology: A Multidisciplinary Approach*. He is a fellow of the American Association for the Advancement of Science and of the American Geophysical Union, which awarded him the James B. Macelwane Medal. Other awards include the Harold C. Urey Prize (American Astronomical Society) and Ya. B. Zeldovich Award of COSPAR's Commission B. Dr. Lunine serves on the Space Science Advisory Committee and chairs the Solar System Exploration Subcommittee for NASA. He earned a BS in physics and astronomy from the University of Rochester in 1980, followed by MS (1983) and PhD (1985) degrees in planetary science from the California Institute of Technology.

Christopher P. McKay, NASA Ames Research Center. Dr. McKay received his PhD in astrogeophysics from the University of Colorado in 1982 and has been a research scientist with the NASA Ames Research Center since that time. His current research focuses on the evolution of the Solar System and the origin of life. He is also actively involved in planning for future Mars missions including human exploration. Dr. McKay has been involved in research in Mars-like environments on Earth, traveling to the Antarctic dry valleys, Siberia, the Canadian Arctic, and the Atacama Desert to study life in these Mars-like environments. He was a co-I on the Titan *Huygens* probe in 2005, the Mars *Phoenix* lander mission in 2008, and the Mars Science Laboratory mission in 2012.

Erika Offerdahl, North Dakota State University. Dr. Offerdahl received her PhD in biochemistry from the University of Arizona in 2008. She is currently an assistant professor at North Dakota State University, where her research focuses broadly on the mechanisms by which scientific expertise is developed in undergraduate college students. Specifically, she is interested in the role and use of visualizations in the development of visual literacy in the molecular life sciences and the effective implementation of assessment in developing students' understanding of the skills and disciplinary practices of scientists. Offerdahl is co-author of an undergraduate astrobiology lab activities manual and is a National Academies of Science Teaching Fellow in the Life Sciences.

Ted Peters, Institute for Theology and Ethics, Pacific Lutheran Theological Seminary, and the Graduate Theological Union. Dr. Peters is a professor of systematic theology at Pacific Lutheran Theological Seminary and the Graduate Theological Union. Along with Robert John Russell, he co-edits the journal *Theology and Science* at the Center for Theology and the Natural Sciences. He is the author of *The Evolution of Terrestrial and Extraterrestrial Life* (Pandora Press, 2008) and co-author with Karen Lebacqz and Gaymon Bennett of *Sacred Cells? Why Christians Should Support Stem Cell Research* (Roman and Littlefield, 2008). Along with Martinez Hewlett, he is co-author of *Can You Believe in God and Evolution?* (Abingdon, 2006).

Margaret S. Race, SETI Institute. Dr. Margaret Race is an ecologist at the SETI Institute in Mountain View, California. Her professional interests focus on the environmental and policy implications of diverse science and technology projects. Her main research focus involves working with NASA and the international scientific and space communities in developing, refining, and applying planetary protection policies and regulations to space exploration missions. Over the past decade she has served on numerous National Research Council studies of forward and back contamination associated with missions to Mars and other celestial bodies. She also served as an organizer and editor of a series of international workshops on the planetary protection implications of both Mars sample return missions and human missions to Mars. In addition to her research and analytical work, Dr. Race is actively involved in science education and public outreach about astrobiology through the mass media, schools, museums, and presentations for general audiences. Dr. Race received a BA degree in biology and an MS in energy management and policy from the University of Pennsylvania and a PhD in ecology from the University of California at Berkeley.

Woodruff "Woody" T. Sullivan III, University of Washington. Dr. Sullivan's interests are in astrobiology (life in a cosmic context) and ways in which astronomy relates to the broader culture. He is a professor of astronomy and an adjunct professor of history at the University of Washington, where until recently he was chair of the steering group of a pioneering interdisciplinary graduate program in astrobiology. Together with John Baross he edited the first graduate-level textbook in astrobiology, *Planets and Life: The Emerging Science of Astrobiology* (Cambridge University Press, 2007). He has long been active in the search for extraterrestrial intelligence (SETI), and was one of the founders of the seti@home project, which has involved more than 5 million participants on the Web since its launch in 1999. His research on the history of astronomy has emphasized the twentieth century; his recent book is *Cosmic Noise: A History of Early Radio Astronomy* (Cambridge University Press, 2009). Sundials are a special and peculiar passion. He has designed many sundials, one of which (a Mars-Dial, the first extraterrestrial sundial) landed on Mars in 2004 as part of the Mars Exploration Rovers. Dr. Sullivan received his BS in physics in 1966 from MIT and his PhD in astronomy in 1971 from the University of Maryland.

Jill Cornell Tarter, SETI Institute. Dr. Tarter holds the Bernard M. Oliver Chair for SETI (Search for Extraterrestrial Intelligence) and is director of the Center for SETI Research at the SETI Institute. Dr. Tarter received her Bachelor of Engineering Physics Degree with Distinction from Cornell University and her MS and PhD in astronomy from the University of California, Berkeley. She served as project scientist for NASA's SETI program, the High Resolution Microwave Survey, and has conducted numerous observational programs at radio observatories worldwide. Since the termination of funding for NASA's SETI program in 1993, she has served in a leadership role to secure private funding to continue the exploratory science. Currently she serves on the management board for the Allen Telescope Array, a joint project between the SETI Institute and the UC Berkeley Radio Astronomy Laboratory. Tarter's work has brought her wide recognition in the scientific community. In 2004 *Time* magazine named her one of the Time 100 most influential people in the world. Tarter is deeply involved in the education of future citizens and scientists. In 2009 Tarter was awarded a TED Prize for her leadership in her field, her unconventional viewpoint, and her vision to change the world. Many people are now familiar with her work as portrayed by Jodie Foster in the movie *Contact*.

Mark Triant describes himself as "a peregrine philosopher and creative technologist." While an undergraduate at Reed College, he worked with Professor Mark Bedau and co-authored the paper "Social and Ethical Considerations of Creating Protocells," which appears in this volume in updated form. He received a BA in philosophy from Reed College, where he wrote a senior thesis entitled "The Problem with Realism: A Case Study in the Limits of Inquiry." He received a master's degree from the Interactive Telecommunications Program at NYU, where he studied prosthetic and environmental approaches to perceptual augmentation. He currently works in the private sector in Brooklyn, New York, researching the future of the infrastructure of human knowledge.

Elspeth M. Wilson, University of Pennsylvania. Elspeth Wilson is a doctoral candidate in political science at the University of Pennsylvania, where she is writing a dissertation titled "The Reproduction of Citizenship." In addition to being awarded Benjamin Franklin and Presidential Prize Fellowships to fund her graduate studies at the University of Pennsylvania, Wilson served as the administrator for the Penn-Mellon Foundation Program on Democracy, Citizenship, and Constitutionalism in 2007–2011. She was also the recipient of a Teece Dissertation Research Fellowship for the summer of 2012, awarded by the School of Arts and Sciences to fund interdisciplinary empirical field research by graduate students in the sciences or social sciences, and she was the Penn-Mellon DCC Program's Boise Family Graduate Student Fellow for the 2012–2013 academic year. Wilson received her MA from the University of Wisconsin at Madison, and her BA with honors in political science from Columbia University. Her research interests include ethics and public policy, democratic theory, American political development, public health and social welfare, political and legal philosophy, civil rights, and constitutional law.

Neville J. Woolf, Emeritus Professor of Astronomy at Steward Observatory of the University of Arizona. In 2003 the College of Science of the University of Arizona named Dr. Woolf a Galileo Circle Fellow. In his career Woolf has received many other fellowships and awards, including a Fulbright Scholarship and an Alfred P. Sloan Fellowship. He is best known for the identification of silicates as being ejected from red giant stars, appearing in the interstellar medium, and present in comets. He also developed the rationale for building telescopes with lightweight mirrors in thermal equilibrium with the air. With Roger Angel he helped initiate designs of telescopes and interferometers for observing terrestrial planets around other stars. In addition to his scientific achievements, he has served on committees related to finding terrestrial planets and astrobiology, such as the Terrestrial Planet Finder (TPF) Science Working Group (SWG) (2000–present); TPFSWG Biomarkers Subgroup, Astrobiology Roadmap Revision, National Academy Committee on Origin and Evolution of Life (2003–2006); National Academy Study of the Scientific Context for the Exploration of the Moon (2006–2007); and NASA TPF Interferometer Science Working Group (2005–2006). He was the principal investigator of the Tucson node of NASA's Astrobiology Institute (2003–2008). He has published over 135 scientific articles. Woolf obtained a BSc in physics (1956) from Manchester University and his PhD in astrophysics (1959) from Manchester University.

Editors

Chris Impey, University of Arizona. Dr. Impey is University Distinguished Professor at the University of Arizona. As deputy head and academic head of the Department of Astronomy, he runs the nation's largest undergraduate majors program in astronomy and the second largest PhD program. His research interests center in observational cosmology, gravitational lensing, and the evolution and structure of galaxies. As a professor he has taught astronomy to more than 4,500 students and has won 11 teaching awards at the University of Arizona. He has pioneered curriculum development in astrobiology and was the principal investigator on a major four-year grant from the Templeton Foundation that explored issues at the interface of science and religion. Working with planetary scientist Bill Hartmann, he co-authored two introductory textbooks that have sold over 100,000 copies. He has written more than 30 popular articles on cosmology and astrobiology. He is the creator of the Teach Astronomy website, which supports nonscience majors, and he's taught parts of his classes in the 3D virtual world called Second Life. He has an Internet startup called "Student on the Go." In 2001 he was named University Distinguished Professor; in 2005 he was honored as Galileo Circle Scholar, the College of Science's highest honor. The Carnegie Council for the Improvement of Teaching named him Arizona Professor of the Year in 2005, and he was one of only six people nationwide named Distinguished Teaching Scholar by the National Science Foundation. In 2007 Random House published his first popular science book, *The Living Cosmos*, on astrobiology. His second and third popular books, *How It Ends* and *How It Began*, were published in 2011 and 2012 by Norton. He has published more than 160 refereed articles. He obtained his PhD from the University of Edinburgh in 1981.

Anna H. Spitz, OSIRIS-REx Asteroid Sample Return Mission Education and Public Outreach Lead. Since joining University of Arizona in 2000, Dr. Spitz has worked on various projects, including work with the Center for Astrobiology, Biosphere 2, and SkyCenter. At the Center for Astrobiology, she worked with Director Jonathan Lunine and the steering committee to establish interdisciplinary astrobiology minors, to host workshops, and to raise funds for research and activities through grant proposals. She worked in the environmental field for over a decade before joining the University of Arizona. Spitz serves on local boards and continues to participate in research and consulting in the environmental field as well as freelance writing activities. Spitz received an AB degree from Wellesley College in 1976 and an MBA from Duke University in 1978. She returned to school and received an MA in geological sciences in 1985 from SUNY Binghamton and in 1991 obtained a PhD in geosciences from the University of Arizona for her work on meteorites.

William R. Stoeger, Vatican Observatory. Dr. Stoeger is a senior staff scientist for the Vatican Observatory Research Group at the University of Arizona, Tucson, specializing in theoretical cosmology, gravitational physics, and interdisciplinary studies bridging the natural sciences, philosophy, and theology. He is a Jesuit priest with an AB degree in philosophy and physics from Spring Hill College in Alabama, an MS in physics from UCLA, and an STM degree in theology from the Jesuit School of Theology in Berkeley, California. After completing his PhD in theoretical astrophysics at

Cambridge University and postdoctoral research in gravitational physics at the University of Maryland, Stoeger joined the Vatican Observatory staff in 1979. Besides his contributions to black-hole astrophysics and cosmology, he has been very active in organizing and participating in research initiatives exploring the interface between science and theology, and science and philosophy, and contributing many articles to volumes and journals dedicated to those subjects, most notably the Vatican Observatory/CTNS workshops and series *Scientific Perspectives on Divine Action*. For a number of years he taught a popular course, Science and Theology, with Tom Lindell in the Molecular and Cellular Biology Department at the University of Arizona.

Index